DESIGN FOR SECURE
RESIDENTIAL ENVIRONMENTS

DESIGN FOR SECURE RESIDENTIAL ENVIRONMENTS

A Technical Handbook by
STEVE CROUCH, HENRY SHAFTOE and ROY FLEMING

LONGMAN

The CHARTERED
INSTITUTE OF
BUILDING

Addison Wesley Longman Limited
Edinburgh Gate
Harlow, Essex CM20 2JE, England
and Associated Companies throughout the world.

Co-published with The Chartered Institute of Building through
Englemere Limited
The White House, Englemere, Kings Ride, Ascot
Berkshire SL5 8BJ, England

First published 1999

ISBN 0 582 27660-8

British Library Cataloguing-in-Publication Data

A catalogue record for this book is
available from the British Library.

Set by 35 in 10/12pt Ehrhardt

Transferred to digital print on demand 2002

Printed and bound by Antony Rowe Ltd, Eastbourne

CONTENTS

ACKNOWLEDGEMENTS

The authors wish to thank Holophane Europe and Designplan Lighting Ltd whose support enabled the research for the chapter on lighting to be carried out, and the Department of the Environment who originally sponsored the work for the building detailing section.

Thanks to Paul Stollard for sowing the seeds of this book, and to Steve Osborn and Heather Graham of the Safe Neighbourhoods Unit for assistance with production.

CHAPTER 1

INTRODUCTION

This book considers the potential for reducing crime and improving community safety through good practice in the construction or rehabilitation of buildings and their environments. Guidance is provided on the processes of: assessment, design, specification, detailing, installation and maintenance of security features. While every building and neighbourhood has different problems and requires a unique design solution, it is possible to make good decisions based on a proven range of design and security features. This book does not offer a rigid set of rules but will nevertheless give detailed guidance and specifications for the design of vulnerable parts of dwellings and their environment.

This book deals with physical security measures for dwellings, community buildings (such as community centres and neighbourhood offices) and small commercial premises (such as shops and workspaces). It does not cover the security of large buildings or conglomerations such as factories, office blocks, and shopping malls.

As levels of crime and fear continue to be a major cause for concern both in Britain and abroad, there is a need for sensible guidance for built environment practitioners who are involved in the design, construction, maintenance or refurbishment of all types of small buildings, regardless of current levels of risk in the neighbourhood concerned.

1.1 PRINCIPLES OF SECURITY DESIGN

Good design for secure residential environment entails a measured and balanced approach. We want the security of buildings to be based on good sense rather than the exhortations of hardware manufacturers or simplistic concepts of fortification. Equally, it is important that the occupants and users of buildings feel safe within them, as fear of crime can often be as debilitating a state as the experience of crime itself.

1

If we merely wished to make buildings as physically secure as possible (and we had the funds to do so) we would probably all end up occupying spaces similar to bank vaults. We would become depressed from lack of natural lighting (windows are far too vulnerable as points of entry) and we would suffer from claustrophobia. In other words, security would be achieved at the expense of a decent quality of living.

What we need to achieve through good design is an appropriate range of security features that reconciles the reduction of crime risk with the human need to occupy pleasant surroundings – where one has a sense of security without oppressive fortification. In the commercial sector, the redesign of the interiors of the major clearing banks is a good example. Twenty years ago you spoke to the bank teller through a slit in a barred widow and your cash was passed to you via a sliding bucket; today most banks feel as though you have wandered into the manager's study, with only the discreetest of physical barriers between customer and staff, and yet in reality bank security is better than ever.

1.2 POTENTIAL AND LIMITATIONS OF PHYSICAL SECURITY MEASURES

Physical security measures aim to reduce the number of opportunities to commit crime. Many property offences occur because someone sees the chance to gain access to a building or room easily and get away unseen. It is clearly sensible to reduce the number of easy opportunities to commit offences. But in the broader offending picture there are two intervening complications: 'raising the stakes' and displacement. As you secure more buildings, many opportunistic offenders will just give up. However, others will become more desperate and some will see increased security as a challenge to their ingenuity. In a different crime context, *Which* magazine reports on new cars show that, despite manufacturers' attempts to secure their latest models, there are still only a handful of cars that cannot be broken into by a professional thief in a matter of seconds. Offenders such as drug addicts and recidivists who 'need' to go on offending may use more force or determination when confronted with increased security barring their way to the items they seek. Bored young delinquents may perceive a hardened target as a demanding game to be won. Persistent offenders, when confronted with a more secure barrier to their goal, may displace their attention to other, more vulnerable properties or they may change the type of offending they indulge in. There have been cases where, after a neighbourhood security programme has been completed, burglary has decreased but street robberies have increased.

It should be clear from all of the above that although physical security measures have an important role to play in increasing community safety, they are unlikely to provide a total solution to the problem of crime in a particular neighbourhood. Ideally they should constitute part of a package that will also include social and management measures.

1.3 COMPREHENSIVE SECURITY DESIGN STRATEGIES FOR WHOLE BUILDINGS AND THEIR ENVIRONMENTS

Just as physical security should be part of the jigsaw that will include social facilities (e.g. play and youth provision) and management approaches (e.g. community policing), so individual buidings and premises should be seen as parts of a secure whole – i.e. the block or neighbourhood. A building is only as secure as its weakest point and an isolated or concealed location will permit the offender to get away with the use of more force to gain access. Therefore a stock solution, such as the upgrading of all front doors in a block or street, will be ineffective unless a risk assessment survey has shown that this is the most vulnerable point of entry and that all premises share an equal risk of being broken into – a most unlikely scenario.

On average, front doors are only used as the entry point in one-fifth of all burglaries – sides and backs of buildings are usually much more vulnerable (see Diagram 1).

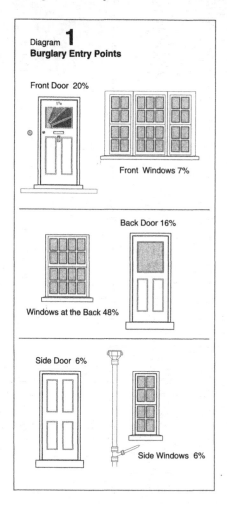

Diagram **1**
Burglary Entry Points

Front Door 20%

Front Windows 7%

Back Door 16%

Windows at the Back 48%

Side Door 6%

Side Windows 6%

Plate 1   Conflict between security and fire safety – Birmingham

1.4 RECONCILING PHYSICAL SECURITY MEASURES WITH FIRE SAFETY

If you make it more difficult for people to get into your property, you run the risk of also excluding the emergency services who may have to force a legitimate entry in the case of fire or serious illness. A number of people have died in recent years when the fire services have been unable to break into their burning premises which had been heavily secured against burglars. Fire escape staircases present particular problems as they often provide concealed access and exit routes for offenders.

Add-on external fire escapes are a typical feature of flat conversions to buildings of three or more storeys. Unfortunately they provide easy access to the upper floor windows, particularly as they tend to be located at the side or rear.

Generally fire escapes are designed in one of two ways:

1. Those which are in the form of ladders, fixed vertically against an external wall. For additional safety precautions, ladders are usually surrounded by a metal cage and small landings are provided on each floor. To guard against both youngsters playing on escapes and burglars, the ladder between ground level and the first landing can be suspended with guide rails between the first and second floor landings. The ladder should be held in place by a simple catch which, in times of emergency, can be easily released thereby lowering the ladder. This type of fire escape, however, is not suitable for use by the elderly, disabled or children.

2. Alternatively, fire escapes can be in the form of a flight of metal stairs. This design undoubtedly takes up more space than the fire escapes described above and for this reason are usually located at the rear of buildings. These can present a greater risk as they are permanent fixtures allowing continuous access to doors and windows.

All windows which are accessible from fire escapes need to be protected. Because of their vulnerable position this usually means extra protection over and above fitting window locks. It may be necessary to reinforce or replace glazing with laminated glass or even to fit window grilles.

Emergency exit doors leading onto fire escapes are also more vulnerable than normal entrance doors. To facilitate ease of escape they are usually required to open outwards and from inside the dwelling they must be opened without the need for a key. Emergency exit doors should be installed to the same specification as that described for external entrance doors to flats. Particular attention should be paid to the hinges because, on outward-opening doors, hinge knuckles are exposed and susceptible to tampering (knuckles sawn off or hinge pins removed). Furthermore, external access to the door release, by way of vulnerable glazing or flimsy timber panels, should be prevented.

For both safety and security purposes, adequate external lighting of fire escapes should be provided.

Any substantial proposals for increasing the security of buildings should be checked with fire safety officers. For more detail on this important issue, see sections 5.0, 5.6 and 6.9.

CHAPTER 2

ASSESSMENT PROCESS

Although there are some general design and specification principles that can be applied to all sites, the 'experts' on security risk and improvements needed in any particular neighbourhood are the people who live or work there. This obviously presents a problem if you are developing a greenfield site, but even here there are likely to be surrounding neighbourhoods where problems can be gauged, or you could consult a panel of intended occupants of your new development.

Many professionals fight shy of consulting laypeople about the work they propose to do, and there seem to be a number of rationalisations for this state of affairs:

1. The professionals know best; ordinary people don't understand the problems and technical considerations.
2. It's obvious what needs doing; there's no need to ask anyone else.
3. Residents might criticise the proposals made by the professionals; this is humiliating and demoralising.
4. Consulting local people and workers is a cumbersome and time-consuming process, which can cause unnecessary delay in getting the work started.
5. Most professionals have not been trained to ask ordinary people for their views on professional and technical matters.

In most cases these are not good enough excuses for proceeding without a consultation stage. The professionals' tendency to keep the client or consumer at arm's length may make for an easy life, but can lead to a waste of resources and inappropriate work. Consulting people requires some expertise and can also be time-consuming, but with proper training and planning the benefits will reward the small extra effort. Local consultation at the planning stage can also be integrated with a 'before' stage of evaluating the work's effectiveness – an essential ingredient of any rolling programme of improvements (see section 2.4). Failure to consult may lead to a waste of money or effort by securing buildings inappropriately. Local people will know which parts of their immediate environment are least secure (e.g. loft hatches at the top of common stairs or windows accessible from lean-to extensions). They will also tell the designer or surveyor which security devices they are likely to make use of and which would be a hindrance to their daily lives.

2.1 CONSULTATION AND RESEARCH

There are a number of tried-and-tested methods for consulting local people and workers about problems and needs. It is not within the scope of this book to give a step-by-step guide to research techniques, but the general principles, which are good for many types of needs assessment, can be outlined.

The two most likely methods for consultation on security would be questionnaire surveys or meetings.

Questionnaire surveys

A well-structured questionnaire administered to all affected residents is the most rigorous method for acquiring data and views on security matters. Exhaustive, 100 per cent surveys are rarely feasible or necessary, so it is generally quite adequate to interview a representative sample of the population. Numerous security surveys have been undertaken on this basis and generally involve a mixture of qualitative and quantitative questions (e.g. 'Do you think lighting is adequate in your neighbourhood? If "No", . . .'; 'Is lighting a problem in the following areas: . . . ?'; 'Why do you think lighting is a problem in these areas: not enough lights, lights too dull, poor maintenance, vandalised, badly positioned, unreliable time switches?; etc.).

If possible such questionnaires should be completed by a face-to-face interview with the respondent. This ensures a better response rate, less misunderstanding and the potential for teasing out extra qualitative data. Postal surveys are easier to administer, but tend to produce a very low response (usually less than 10 per cent of all questionnaire recipients) and are liable to have a number of questions incorrectly answered.

The respondents to a questionnaire would normally be a random sample of adults to match the actual composition of the local population, but it may sometimes be appropriate to boost the number of respondents in particular categories seen as being most affected by crime. This might include women, elderly people or people from certain ethnic minorities. It will usually also be appropriate to survey key workers such as police and caretakers, using the household questionnaire or an adaptation of it.

The questions asked should be designed to inform the intended outcome – in other words 'What do you want the survey to help you achieve?'. This entails working backwards from an option scenario rather than just throwing together some questions about security in the hope that they will lead to some ideas. If this 'reverse framing' approach is taken it will be much easier to move on to the stage of prioritising options.

Meetings

Sometimes a full questionnaire survey will be neither practicable nor appropriate – for example, when the improvements are fairly modest or the range of options for improvement is limited by external factors. In such cases it is still important to consult with local people, if only to gain their support and co-operation. Meetings have the advantage of being speedier, less labour intensive and allow for 'bulk handling' of opinion. However, this latter point can reduce their effectiveness or representativeness.

Large public meetings, where the experts tell the masses what they have decided to do, are rarely useful in terms of consultation or even public relations. Often they are used as heckling opportunities by vociferous individuals who do not necessarily represent majority opinion. They can lead to heated exchanges and the raising of hostility. It is usually much better to call a series of representative (sampled as for questionnaire surveys) small group meetings or to make a presentation to existing residents' or residents' group meeting. Even in these smaller meetings it is essential to structure the content and process, otherwise the consultant will find that he or she is being harangued about the shortcomings of every other council service in the area!

2.2　PHYSICAL SURVEYS

These are essential before proposing changes to existing developments and may involve surveys of (a) any existing defensive measures, including their condition and use; (b) the external community facilities such as playgrounds, roads, gardens, footpaths, fencing, parking areas, garages; (c) the internal common areas of blocks and communal services such as lifts, entryphones, lighting, refuse disposal and means of escape; (d) the structure and fabric of blocks and houses; and (e) the condition of individual dwellings, including internal spaces and internal services (plumbing, heating and electricity).

Observation of circulation patterns and use of communal and public areas is an important precursor to planning any significant changes of layout. Finally, wear and tear on existing measures (such as entryphones or door closers) and misuse of existing fittings (such as fire exits wedged open or padlocked, or keys left in window locks) can all build up a helpful picture of what is likely to be effective when improvements are made.

2.3　PROPOSALS, PRIORITIES AND OPTION APPRAISAL

If, as suggested earlier, the questions asked in the consultation stage were designed to inform one or more scenarios (equivalent to what academic researchers do to 'test a hypothesis'), it should be fairly straightforward to prioritise the options available or generated. However, it is important that the professional involved does not try to bend the responses to fit with his or her preferred option. One way to avoid this bias is to take the concatenated responses back to a small group of professionals and local representatives. Informed by the consultation, this group can work out the best plan of action. Whichever way the responses are turned into improvements, it will be necessary at this stage to introduce other factors, most notably cost restrictions and technical constraints. Juggling all these factors together will not necessarily lead to one obvious plan, and the final course of action will probably have to be a compromise between the designer's technical/financial constraints and the local ideal. Nevertheless, agreed compromises are better than unilateral decisions.

In times of severely limited resources the least painful compromise can be arrived at through cost–benefit analysis, which attempts to put a value on both quantifiable and non-quantifiable factors in a range of scenarios, so that the one with the highest benefit-against-cost ratio can be selected. The technique for applying cost–benefit analysis to building works is fully explained in the Department of the Environment's *Handbook of Estate Improvement*, Part 1 (see Bibliography).

2.4 EVALUATION

Among hard-nosed practitioners the attitude is often: 'Well we've done the work as planned, so it must be okay'. Finances are often so tight that people quite understandably want to spend everything on the actual improvement works, but saving the cost of evaluation can be a false economy in the longer term. The same implementation mistakes may be repeated in subsequent schemes: landlords and owners may continue to invest in measures that are not actually reducing crime and insecurity, and minor adjustments that could make all the difference to the effectiveness of improvements might never be made.

Evaluation has to be thought about at the earliest stages of planning improvements so that baseline information can be obtained. A significant part of evaluation is based on changes over time, both subjective and objective, and not all data can be obtained retrospectively. This is particularly true of public perceptions of safety and security, which after all is the focus of much of the work covered in this book. The ideal evaluative model would consist of surveys before and after improvement work, combined with chronological comparisons of hard data such as recorded crime rates, police call-outs, running costs and repair bills. The aim would be to assess changes in levels of victimisation, perceptions of safety, and expenditure on crime and maintenance. Ideally, these changes would need to be checked against changes in a comparable 'control' area and any intervening variables such as major population, tenure, policing or social changes in the improvement area. This all may sound like a tall order, but represents a modest time investment for a pay-off which can lead to more efficient future schemes, and can be a major lever for obtaining more security improvement funding from a pool of competing bids.

CHAPTER 3

DESIGN PROCESS: EXTERNAL AND COMMUNAL AREAS

Designers and specifiers of security improvements must be aware of both the long-term and short-term consequences of their decisions. In the short term they should ensure that the upgrading process is as simple and as painless as possible. The 'buildability' of the proposals is crucial. How easy is it to construct or install without causing further problems? These principles need to be applied to each aspect of the design.

In the long term they need to consider how the improvements they make will be 'managed'. This will involve designing to reduce the amount of maintenance necessary and to make repair and replacement simple and inexpensive. They must be conscious of the future revenue consequences of their decisions as well as the short-term capital costs.

For external and communal areas the key factor is access control: 'who is allowed where'. In the public domain – adopted streets, parks and common land – anyone can wander, as long as that person appears to behave properly. Within each private dwelling only the householder and his or her family and guests are welcome. A clear barrier needs to be placed between the public and private domains. This usually consists of the garden gate or the front door, but in many contexts there is a 'grey area' between the definitely public and the definitely private – communal or undefined spaces. This is a particular problem in multi-occupied and non-conventional developments such as Radburn and 'village green' layouts. Even in areas that consist of houses-in-gardens-on-streets, there are quite often back lanes, alleys and left-over corners where it is unclear who is allowed to roam or loiter. Generally community safety improvements will aim to reduce these undefined areas, but there is a risk that we end up eliminating all the areas that allow for neighbourly interaction. For example, people today are reluctant to provide benches in public or communal areas for fear that they will attract rowdy youths, vagrants and litter. Yet one of the pleasures of a well-managed neighbourhood is to be able to sit out and watch the world go by or chat to your neighbours in a shared open space.

There are three main categories of access control: passive physical security, psychological measures and active electronics. Physical security and electronics will be covered in more detail later in this book, so let us here consider the contribution that psychology can make to safety and security, particularly in communal areas.

Plate 2 Symbolic barriers to private space – Bristol

Certain design and layout features can be incorporated in residential developments to either signal that a space is not open to all-comers or to ensure that potential offenders feel they are out of place. The most obvious of these set-ups is the crescent or cul-de-sac where anyone in the area will feel surrounded by the possibility of eyes behind net-curtained windows. This arrangement only works if the (expanded) rear aspect of such developments is secure, otherwise all the eyes will be focusing in the wrong direction. Another popular psychological ruse is to have a symbolic entrance to a development. This is increasingly being applied to new infill estates where the main vehicular and pedestrian access routes have a 'hole in the wall' entrance feature and some change of road and footpath surface treatment to mark the transition from fully public to communal areas.

Such psychological signals will only work if the defined neighbourhood is predominantly well maintained and law abiding, otherwise it can work in reverse, as often happens on drab monolithic council estates were the environment exhibits signs of abandonment and lack of care.

Another psychological measure for making developments less popular for would-be offenders, but conducive to neighbourly identification, is to group small sets of dwellings together either physically or symbolically by using colour schemes, distinctive surface treatments or detailing and new fencing and planting. Such 'breaking up' of larger neighbourhoods can offer a more fertile climate for neighbours to take control of 'their' patch or corner. Although there is no guarantee that this community development will happen, it is nevertheless a useful ploy for breaking up the anonymity and dreariness of large developments.

The above-mentioned approaches offer a softer alternative to the harshness of excessive physical fortification and the 'big-brother' intrusiveness of some electronic measures.

3.1   PEDESTRIAN AREAS

The previous suggestion for neighbourly clusters is based on the well-tried principle that communal space usually only 'works' if it is managed by or on behalf of the residents who use it. The pragmatist may argue that, because communal space is potentially so fraught, it is best avoided at all costs. However, as noted previously, communal space is sometimes appropriate and desirable, and there are many cases where it is not possible to convert it into wholly private or public space, even if this were thought to be preferable (e.g. the communal areas within blocks of dwellings or pedestrian access routes within developments). And what about children and young people's need for communal space?

Traditionally the provision of play and leisure areas has been associated with public housing; but some private developers have become involved in what was previously public property, or in the development of 'villages' within the green belt, and they, too, need to consider where youngsters are going to play or 'hang out'. If sites are planned with cul-de-sacs and through roads incorporating 'traffic calming' measures, children will be encouraged to play in the streets, and appropriate measures should be taken to ensure that adequate provision is made so that they will not be encouraged to play around parked cars. Where play areas and pocket parks are provided they should be kept small and be visible from people's homes, though not too close as to cause noise nuisance. As a general rule, play areas for small children should be placed close to dwellings, whereas larger 'kick about' areas for older children and teenagers should be located further away. Problems tend to emerge when older children use or 'hang about' in play areas designed for youngsters and have no specific area of their own.

Well-used, well-lit footpaths overlooked by a number of dwellings or building entrances are the safest for pedestrians and the least popular with offenders. Passing vehicles can also provide casual surveillance of footpaths beside roads and reassurance to pedestrians, providing that some measures are taken to ensure that vehicular speed and volume are reduced. Where access routes do not generate much traffic it is even more important that they are overlooked by the frontages of dwellings.

In new developments unnecessary footpaths, particularly those which do not follow vehicle routes and might provide intruders with unobserved access and escape routes, should be avoided. If the choice of layout makes separate footpaths necessary, they should be kept short, direct and well lit. Long, dark alleyways should be avoided at all costs. When providing access to the rear of terraced property, paths should not join to make a through route, though it is generally best to avoid any rear access where possible.

A high proportion of vandalism can be attributed to people taking short cuts between lines of circulation. Sharp changes in direction should be avoided and, if

Plate 3   Potential conflict between plant growth and sight lines

this is not possible, feeble items such as litter bins or low walls should not be used as markers as they will soon be damaged.

Landscaping and planting associated with footpaths should reinforce security and should not exceed one metre in height where abutting pavements, thus ensuring that it does not obstruct lighting or 'bush' out to form a hiding place.

The most extreme examples of uncontrolled pedestrian access are in estates where blocks of flats or maisonettes are freestanding in open space. Public access should be restricted by enclosing the space around blocks with a communal garden or private gardens for ground-floor dwellings. Where essential routes are neither well-used nor overlooked, as is the case with the internal corridors of linear blocks, it may be necessary to restrict access to residents only.

Walls, fencing and gates

It is not always easy to impose restrictions on the established routes of people within residential areas. Attempts are sometimes made to divert people away from familiar or convenient routes by the use of fences or other barriers. However, these routes can quickly be re-established by damaging the barrier or by using gates provided for emergency access only. Proper consultation with residents and observations of pedestrian movements are essential if any reorganisation of routes and footpaths is to achieve the intended objectives. It is important to note that 'desire lines' may only be established after a period of occupancy and flexibility should be incorporated into the design layout. In addition, the shortest route may not necessarily be the desired route. The design, detailing and construction of any fence or barrier which closes off a short cut or convenient, but unauthorised, route must be carefully considered. Local authorities have usually found it cost-effective to invest in high-quality barriers, as even robust two-metre-high wooden perimeter fencing can start to deteriorate after four to five years – especially prior to November the fifth!

Despite abortive attempts by designers in the 1960s to promote open-plan surroundings to domestic premises, walls and fences have proved to be essential elements of the residential environment, for community safety purposes. Walls and fences have to be designed so that they do not create dark corners or restrict natural surveillance. A high fence or wall does much to keep intruders out but, as others argue, once intruders gain entry they are hidden from view from the street or public area.

This problem is at the root of fence design considerations. It is therefore best to establish a form of fencing that is both difficult to climb but is in part see-through. There are a number of different priorities for fencing and it is a matter of getting the balance right. For example, in many circumstances residents may prefer a high solid fence for privacy. However, where security is paramount the following two methods are effective:

- Establish a firm, low, stub wall of brick or stone, etc., with some form of trellis covered in a growth of thorn-type planting. Not only do the thorns make climbing difficult but the trellis helps the growth and provides an unstable 'fence' which is more difficult to scale than something more solid.
- Use a more standard security type of fencing. For domestic purposes this type of fencing is, on its own, unsightly but, combined with a sensible planting programme, can provide fencing that is difficult to scale with a see-through surveillance opportunity plus a green-screen which actually looks pleasant.

There are, of course, many variations on these two basic notions.

At the front of properties a one-metre-high wall or fence will clearly define the private area, while permitting casual surveillance by neighbours, pedestrians and passing motorists. At the side and rear of private dwellings 1.8-metre-high, close board fencing will provide privacy and the enclosure of a safe play area while restricting access to these more vulnerable areas. Stronger fencing (e.g. feather edge) will be needed between a private garden and a public footpath than between individual gardens where interwoven fencing would do.

Attention should also be paid to the type of gate specified. A wrought iron gate is suitable where good surveillance is required, whereas solid timber gates offer more privacy to the occupant (and also to the intruder!). All gates should be fitted with at least one robust gate bolt and be hung on a minimum of three hinges (hooks and bands, as opposed to flimsy 'T' hinges).

Any features made of wood are less resistant to the elements and need regular maintenance. They are also more vulnerable to attack by vandals: screws and nails can be pulled out, and wood burns or breaks more easily than other materials.

Plate 4   Radburn layout – South Wales

Plate 5   Modified Radburn layout – Cannock

3.2   CAR PARKING

One of the biggest crime categories is theft of and from cars – which is not surprising when one considers that each car and its contents consists of several thousand pounds' worth of private property and is left unattended for the majority of the day and night. The risk becomes substantial when this mobile piece of property is left on a public thoroughfare or is parked away from the eyes and ears of its owner. Parking provision for residents should therefore be located as close as possible to the owner's dwelling. Ideally, car parking should be located in individual garages within the curtilage of the dwelling, with the approach and entrance visible to occupants. The position of garages and car-ports should not obscure the general view.

A radical solution to the problem of car parking and traffic was designed into a number of housing developments in the 1960s and 1970s. This was known as the 'Radburn layout' where cars and pedestrians were totally segregated. Traffic, parking and garages were contained down one side of a terrace or cluster of houses while gardens (often 'open plan') and footpaths were laid out on the other side. This looked good on paper but has turned out to be a disaster from a security point of view. Burglars were able to get right up to house frontages and to escape down the numerous footpaths and alleys while, on the other side, car thieves were carrying out their business unseen. Many of these Radburn estates have been extensively redesigned to return them to the conventional hierarchy of: house – front door – front garden – parking – footpath – street.

Residents of multi-occupied blocks can have severe problems with insecure parking, unless a secure basement car park was built into the original design. Areas of communal off-street parking around blocks should be divided into small groups so that occupiers can become familiar with the cars and their owners, and thereby detect intruders. Unassigned parking spaces should be off the road in small, private, well-lit groups under natural surveillance and as close as possible to the dwellings. These should be blended in with the street rather than screened or obscured behind planting.

Parking compounds and remote garage courts form ideal sheltered spaces in which children can play or vandals can strike, especially since such areas are rarely under the natural surveillance of the buildings they serve. It is generally agreed that distant spaces such as these should be avoided, as should the provision of parking next to footpaths. However, some sources argue that where both the parking component and the entrance are situated under natural surveillance from neighbouring dwellings, then the fact that this is private space, not to be used as a play area, becomes apparent. Wherever situated, all garage doors should have strong bolts and locks.

Unsupervised, covered or multi-storey car parks have often proved unacceptable to residents and many have been demolished. Although many attempts to provide technological security systems to covered parking areas have been a failure, there are some examples where individual lock-up garages and the provision of a single controlled entrance to communal garages, using 'roll-over' gate access systems, have worked for several years, even in relatively high-risk areas.

3.3 COMMUNAL SPACE WITHIN BUILDINGS

Housing authorities have tried a variety of design and technical solutions to the problems of safety in communal areas of multi-occupied dwellings. Some improvement schemes in tower blocks have created just one access point, opening into a supervised lobby. Others in linear blocks with staircase access have increased the number of access points and restricted the number of people using each one, constructing new entrance lobbies with stairwells serving only a few dwellings. Alternatively, access to dwellings can be restricted to one access point by a simple 'zone-locking system'. Access to landings and balconies for non-residents is prevented by locked gates or doors designed for access/escape in emergencies only.

The use of common entrances should be avoided wherever possible and individual flat entrances should ideally be located at ground level. Where communal entrances are unavoidable – for example, in multi-occupancy flats and maisonettes – they should be arranged so that as few residents as possible use each entrance. This enables residents to get to know each other quickly and develop natural protection

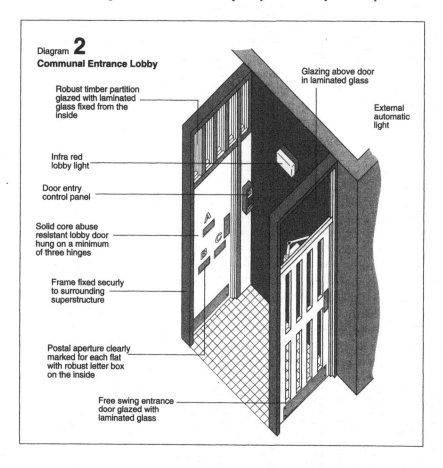

Diagram **2**
Communal Entrance Lobby

Glazing above door in laminated glass

Robust timber partition glazed with laminated glass fixed from the inside

External automatic light

Infra red lobby light

Door entry control panel

Solid core abuse resistant lobby door hung on a minimum of three hinges

Frame fixed securely to surrounding superstructure

Postal aperture clearly marked for each flat with robust letter box on the inside

Free swing entrance door glazed with laminated glass

(the principles of natural surveillance and actual neighbourhoods). The security of communal entrances can be enhanced by installing some form of door entry or access control system. Where a small number of dwelling are served by a communal entrance (up to about six), a simple form of entry system can be used as shown in Diagram 2. These systems are are relatively inexpensive to install and to maintain as all of the components used to make up the system are readily available from electrical suppliers. These systems are not, however, particularly robust and should not be specified where communal entrances serve a greater number of dwellings, such as a multi-storey block. In these situations, stronger and more durable systems are required which will withstand much more wear and tear and, to a certain extent, abuse. (These systems are covered in more detail in Chapter 6.) Communal entrances should always be discrete, obviously for residents only and not through ways providing short cuts.

Problems of security and crime are often exacerbated where there are long access balconies and interconnecting walkways. Shortening the effective length of these access routes reduces the number of people using each entrance and may allow the successful introduction of entry phone systems. This can be achieved by demolishing walkways or by erecting permanent barriers. The horizontal or vertical 'partitioning' of blocks can create zones into which access is only possible with proximity cards or 'suited' keys. However, this is only likely to work if there are exceptionally good caretaking and management arrangements.

Any shared facilities which are provided – for example, drying areas or rubbish collection facilities – should be located within a secure area and in view of the dwellings they serve in order that they are protected by natural surveillance. Well-lit, vandal-resistant surfaces should be provided in this secure private area.

3.4   COMMUNAL ENTRANCES

Communal entrance designs for houses converted into flats vary widely. Two communal entrance designs are described here. The first design, shown in Diagram 2, involves the creation of an internal lobby with an outer lobby door, free swinging and unsecured, and a second secured inner door. The second design has no lobby. The communal entrance to the second and third floor flats is through a single front door, controlled by a door entry system. Entry to the ground floor flat is via a side passageway, as shown in Diagram 3.

Communal entrance lobbies are areas of semi-public space, readily accessible to strangers and open to abuse. To discourage abuse, the outer lobby door should incorporate a large area of glazing to make potential miscreants easily visible from outside. This is best achieved by installing a door with a number of narrow vertical strips of laminated glass, fixed from the inside of the lobby with glazing beads held in position by screws (see Diagram 4). The restricted widths of glass make it more difficult to break and less of a target for vandals. All the glazing in the communal entrance lobby can be dealt with in this manner.

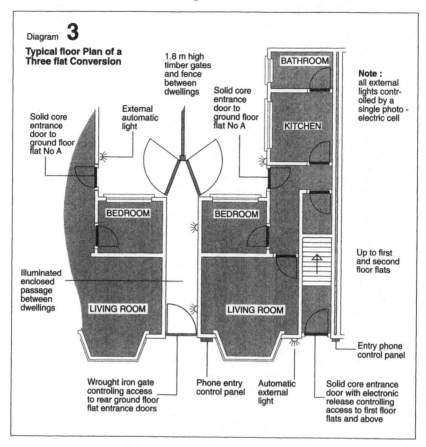

Diagram **3**

Typical floor Plan of a Three flat Conversion

1.8 m high timber gates and fence between dwellings

Solid core entrance door to ground floor flat No A

BATHROOM

Note : all external lights contr- olled by a single photo - electric cell

External automatic light

Solid core entrance door to ground floor flat No A

KITCHEN

BEDROOM

BEDROOM

Up to first and second floor flats

Illuminated enclosed passage between dwellings

LIVING ROOM

LIVING ROOM

Entry phone control panel

Wrought iron gate controling access to rear ground floor flat entrance doors

Phone entry control panel

Automatic external light

Solid core entrance door with electronic release controlling access to first floor flats and above

Adequate lighting should be provided within the lobby to aid surveillance during the hours of darkness. A movement detector is an effective method for controlling lobby lighting, activating the light when someone enters the lobby and switching off after a short period of time. This has the advantage of being economical to run as well as acting as an indicator when there is a someone in the lobby.

The internal lobby door or, where there is no lobby, the single communal door, needs to be effectively secured. Where this door is connected to a door entry system, it must be sufficiently robust to withstand constant use and possible abuse. Failure of a door can invariably be attributed either to flimsy materials used in its construction or to the inadequate fixing together of door components. To reduce the chances of failure:

- The door frame should be secured to the surrounding structure with a minimum of four frame-fixing screws or, preferably, with bolts in each of the two door frame jambs. The door stop should be at least 18 mm, although a 25 mm stop is

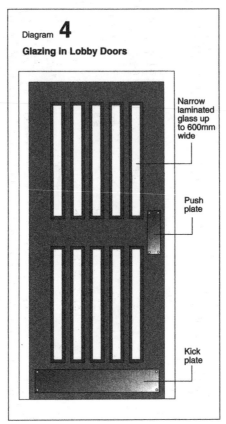

Diagram **4**

Glazing in Lobby Doors

Narrow
laminated
glass up
to 600mm
wide

Push
plate

Kick
plate

preferred. The stop can be either cut from the solid timber or be 'planted' onto
the frame, fixed with both glue and screws.

- The door should be of solid-core construction manufactured from robust
  materials. It should be devoid of weak timber panels and designed to resist force
  and leverage.
- The door should be hung on a minimum of three 100 mm steel hinges fixed
  securely to both the door and door frame. At least two hinge bolts should be fitted
  to reinforce the hinges, fixed approximately 50 mm above the bottom and middle
  hinges. A third hinge bolt may be necessary on particularly vulnerable doors, and
  should be fixed about 50 mm below the top hinge.
- The frame surrounding the door should not incorporate vulnerable glazing or
  flimsy timber panels which can be easily broken to gain access to the door release
  mechanisms.

Careful consideration needs to be given to mail deliveries. Timber post boxes
have sometimes been mounted on an external wall close to the communal entrance
door, but these exposed boxes have frequently proved to be inadequate in
preventing the theft of mail. They can also be easy targets for vandals.

No more than three postal apertures should be provided through the secured communal entrance door. An internal robust and lockable post box is need for each aperture, and should be:

- fixed securely on to the surface of the door;
- deep enough to prevent mail from being stolen;
- designed to prevent outside manipulation of the lock through the aperture.

Wire baskets do not provide privacy for mail and should not be used on communal doors.

Where three or more flats are involved, additional post boxes can be provided in the entrance lobby. As these boxes are more exposed they need to be made from strong materials such as steel and fixed securely to a solid wall. Once again they need to be deep enough to prevent the theft of mail. An example of a robust mail box is shown in Diagram 5.

To avoid confusion, postal apertures and post boxes should be clearly marked for each individual flat. Permanent markings can be best achieved by engraving the letter flap or box with the flat number or letter. Flimsy plastic or alloy numerals/letters are easy targets for vandals and should be avoided.

3.5 DESIGN AND LAYOUT OF ENTRANCE FOYERS

In order to accommodate both the staff and the equipment for new access arrangements to multi-storey blocks, it may necessary to redesign the area inside the main entrance. Although the actual layout will depend on the circumstances of a particular block, the following considerations should be incorporated:

- The foyer and reception area should look welcoming. The area could contain, for example, seating, notice boards, plants and pictures.
- The staff area should be secure without giving the impression of a fortified enclave. This is usually achieved by positioning a fairly high counter between staff and visitors/residents. Plate glass screens (as used in banks and DSS offices) should be avoided as far as possible.
- Provision should be made to completely secure the staff/equipment area for times when nobody is on duty, by, for example, a pull-down roller blind.
- The staff control area should ideally have a direct view of both the area outside the main entrance and the foyer. Slightly raising part of the staff area floor level can facilitate this.
- In high-risk areas it may be necessary to install an under-the-counter 'panic button' with a direct link to emergency services.
- CCTV monitors and access control equipment should be positioned so that they are not directly visible from the foyer area.
- Staff will need to have access to refreshment and toilet facilities.

In many cases the additional space for the reception office and equipment has to be found by adapting an adjacent storeroom or ground floor flat.

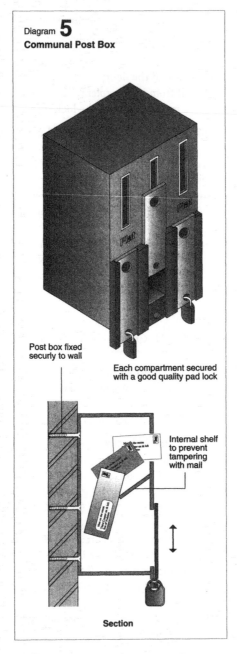

Diagram **5**
Communal Post Box

Post box fixed
securly to wall

Each compartment secured
with a good quality pad lock

Internal shelf
to prevent
tampering
with mail

Section

Plate 6   New reception area for tower blocks – Dagenham

3.6   ACCESS CONTROL AND SECURITY

Nowadays, many entrance areas to multi-storey blocks are kitted out with
sophisticated electronic systems to extend the vision and 'reach' of housing staff and
occupants (covered in detail in Chapter 6). However, it is important to note that
entryphone and CCTV systems are not security systems but access control systems.
They may form part of a security package, if coupled with block receptionists,
television surveillance or strengthened front doors. In most cases, reliance should
not be placed on technological measures alone to improve security. They may be
vandalised or stolen, and will not necessarily reduce crime or make people feel safer.
In multi-storey blocks, the level of security given to individual dwellings should
depend on the level of risk and adequacy of security at the block entrance and in the
communal areas. Correct assessment of the requirements of each block is essential to
identify appropriate individual and collective security measures and the right
combination of design, management and technological approaches.

CHAPTER 4

DESIGN PROCESS: LIGHTING FOR SAFETY AND SECURITY

INTRODUCTION

People have rarely been too happy about the dark unless they are up to mischief. Each 24-hour cycle brings with it a period of fear and uncertainty. Although, for most urban dwellers living in artificially lit environments, the primordial nightly fear of darkness has all but disappeared, it has been replaced by something less predictable, more invidious and apparently worsening – crime.

It is people's perceptions of their environment and what happens in it that is important. But in our age we need figures and statistics, even if we are suspicious of them, to justify our beliefs and actions. It is not until we come to this century that such evidence for the relationship between lighting and crime becomes available.

More must be said later, and in detail, about the scientific analysis of the subject, but the subject only became of real research interest in the 1960s in America when a series of lighting evaluation programmes were carried out by the Federal Law Enforcement Agency. These were followed some 20 years later in Britain by the first European project in Edmonton, London. All these programmes show a definite connection between anti-social behaviour, fear, anxiety and public lighting.

The increasing development and use of the motor car saw a parallel development in the technical development and use of street lighting. No wonder that the development of street lighting, until recently, was geared almost solely to motor traffic. In so doing it set up its own codes of practice, disciplines and nomenclature directed towards the interests of the motorist and the very real problem of road accident prevention.

However, in the last 20 to 30 years developments in the built environment have suggested a new approach to the use of public lighting. Pedestrian precinct areas have flourished, and large areas of housing development are now pedestrian only. Underpasses and footbridges that were relatively few in number, are now common features of new town developments. All these situations demand lighting different to that suitable for traffic.

There is no denying the importance of lighting these areas purely for the purpose of assisting visibility for pedestrian circulation, but the biggest priority has come from concerns about crime and the fear this provokes.

Lighting for traffic and lighting for people are by no means incompatible but they are distinctly different disciplines. If lighting for people (i.e. pedestrians) is to be effective, a new approach needs to be established with its own codes of practice, disciplines and nomenclature.

For many people, blindness is the worst sense loss that can befall us, but loss of sight does not have to be total to raise fears within us. That loss can be partial, such as at night or in poorly lit areas. Indeed, the ability to see only partially can be worse than no sight at all, as shadows, trees moving in the wind and half-seen impressions can do terrible things to our imagination – something film makers have exploited to the full. Whether these fears are real or not is largely irrelevant. If people perceive dangers within their environment then fear will arise and an ensuing loss of confidence will lead to self-imposed restrictions on where and when to go out.

Some of these fears are purely practical in nature. Diminished ability to see means such places as stairways become difficult to negotiate, uneven paving presents problems easily dealt with in daylight. In this way alone our environment makes us vastly more accident-prone at night, particularly when the artificial lighting provided is insensitively installed or low in quality. But, of course, a major concern for many is crime and the perception of it.

What goes wrong

The vast majority of night-time street crimes are committed on the spur of the moment when an easy opportunity is spotted, or when the chance for provocation or retribution is achievable under a veil of darkness. Some 80 per cent of street crimes fall into this category and have nothing to do with the carefully planned jobs of professional criminals. But this is the sort of opportunist behaviour that causes so much fear and anxiety. So, where does lighting as a means of crime prevention come into all this? The answer is that good lighting, when used in conjunction with other measures, has a key role to play in reducing certain types of crime, and, most importantly, reducing fear.

During hours of darkness our sight is diminished and not knowing what lurks around us raises anxiety in all of us. Add to this the belief, fuelled by the media, that crimes such as mugging, rape and general assault are on the increase, and all we want to do is stay at home, safely locked away from the mean streets. In poor lighting under such conditions you have the seed of community decline. But the results of upgrading lighting can be virtually instant. On housing estates in the UK where improved lighting has been implemented, within days people have said that they felt safer, more relaxed, less fearful and were prepared to go out more.

However, it is important to point out that good lighting cannot be the panacea for all neighbourhood crime problems. Lighting will obviously have no effect on crimes that occur in daylight (for example, most burglaries) or inside people's homes (for example, domestic violence and a high proportion of sexual crimes), but good lighting may help to reduce street crimes, auto theft and vandalism, and will reduce the fear of crime even if it does not reduce actual rates of victimisation. This is

Plate 7   Well-lit estate – London

because fear of street crime is often many times greater than the chances of
becoming a victim, particularly among comparatively low-risk groups such as elderly
people. This fear has very tangible consequences – many women and older people
are too afraid to go out alone after dark, and this can have deleterious consequences
on the quality of their lives. Good communal lighting sends out a positive 'signal'
about a neighbourhood, indicating that it is in an area where people can circulate
with confidence and in safety.

4.1   THE PRINCIPLES

When the eye receives light a series of complex things happens and, despite
tremendous advances in understanding over the last few years, a great deal still
needs to be discovered before we know enough to understand completely the
phenomenon we call sight.

Under various lighting conditions a range of human emotions are experienced,
from deep depression to highly excited states. However, it is not just a matter of
'the more light the better'. Very bright lights can cause discomfort through dazzling,
and high-level lighting that only renders part of the colour spectrum (such as ultra-
violet lights, low-pressure sodium and basic fluorescent lights) can create a sense of
harshness and disorientation.

Criminals generally prefer low lighting levels, but so do lovers (think of
candle-lit suppers) and astronomers (there is currently a campaign by astronomers

to reduce what they call 'artificial light pollution', which is ruining their view of the night sky).

Lighting has as much a psychological as a practical effect.

The American lighting evaluation projects that show positive crime reductions indicate a possible link with behaviour. No serious work has yet been done on this issue and it would involve considerable clinical research, but the circumstantial evidence is so strong that the relationship between lighting levels, emotional responses and possible anti-social behaviour cannot be dismissed, and warrants proper investigation.

A less-complicated and more practical aspect to the use of good lighting in public places, as a means of increasing safety, and reducing fear, is the ability to see better. Loss or diminution of a faculty, and particularly a major faculty, in a public area is bound to give rise to anxiety. Not being able to identify faces is of particular concern. Even innocent people can look very frightening under poor lighting conditions. But to this practical aspect of lighting there are two quite distinct concerns: one is lighting for the 'good citizen' simply going about his or her business and, the other, lighting against the would-be offender. Fortunately, the same lighting consideration can be equally effective in dealing with both concerns.

The key factor is recognition. The good citizen wants to be able to see who is approaching while the would-be offender does not want to be recognised. It is deterioration in this recognition that produces the fear and anxiety in the 'goody' and confidence and opportunity in the 'baddy'. So crucial is this factor within the whole issue of public lighting for safety that all light levels and readings should be taken and designed to operate at face height, which is normally around 1.5 metres.

So what kind of lighting and how much?

In principle, the closer the light is to daylight the better. In practice, the whiter the light the better. Or, at least, these two statements hold good for community safety lighting. But there is a difference between the two. Daylight is complex and multi-coloured and constantly changing both in intensity and colour make-up. It is this constant shift and play of light that stimulates the eye rather than tires it and accounts for the fact that, in normal daylight conditions, we do not suffer the sort of eye strain we experience under artificial lighting. But that is not all. The way the light is filtered through the clouds changes its colour make-up and therefore its 'colour-rendering' abilities.

All this is not only natural but has, over millions of years, dictated the way our eyes evolved. It is this kind of lighting with which we feel safe and comfortable.

There are a number of serious problems in overcoming the deficiencies of artificial lighting if measured against the criteria of natural lighting. The ability to provide a constantly changing light source is not, in any practical sense, possible for safety lighting in normal conditions. Indeed, the conditions would not really warrant it even if it were possible. People do not stay for any length of time in a public space at night and therefore are unlikely to derive the benefits afforded by the fluctuations of natural light.

The real issues for community safety lighting lie in the amount, or intensity, of light and the 'colour rendering' of that light.

Intensity of light

Generally speaking the more light you generate, the more it costs to run. However, as we shall see later, this is more than a reasonable cost benefit and is even more reasonable when measured against other safety and crime-preventative measures. As we indicated earlier, lighting problems are not solved merely by increasing the brightness of the light sources. Bright lights can create the problem of glare and shadows – two complications that are very closely interlinked and both depend on the proper placing of lamps and the way light is distributed from them.

A choice: bright lights or even illumination

It has to be appreciated that the stronger any light falls on an object, the stronger is the shadow. In theory, the problem might seem intractable. Firstly, you introduce stronger lighting to improve the situation only to find that you have created a worse problem by creating glare and shadows. You then introduce secondary lighting to illuminate the shadows and so it would appear to go on and on. The trouble is that the human eye reacts to strong lighting by closing down the aperture of the iris, allowing less light into the eyeball. This makes the surrounding shadows even more difficult to see into. Therefore, the aim should be not just to brightly light an area but to start off with the notion of creating an evenness of illumination in which these conflicts are not allowed to arise in the first place.

When, after spending some $12 million on lighting programmes in the USA, the LEA (Law Enforcement Agency) called for reports on the effectiveness of what had been established, the conclusion was drawn that 'in all probability, evenness of illumination was the biggest contributory factor in lighting as a means of crime prevention'. It has proved to be more effective in this respect to have a generally low level of illumination than to have bright areas. Above all other considerations, evenness of illumination is the hallmark of any good lighting for safety scheme.

Colour rendition

Even if evenness of illumination is the most important single factor in lighting and crime prevention, colour rendition runs a very close second. The natural light generated by the sun illuminates all the natural colours found within the range of normal human vision.

In natural light our eyes absorb the range of wave lengths and it is our brain that interprets the waves to something recognisable as colour. What we see in natural white light is a combination of all the colours we see separated (by the prismatic effect of water particles) in a rainbow. However, this is very different from what we get with artificial (electric) light. The range of colour in all but the best of electrical light sources is very limited indeed by comparison to sunlight.

The problem is no longer one of technology but more one of cost. The better the lighting (the fuller the colour range), the more expensive it is to run. Because so much public lighting is required, it must, by necessity, be far cheaper than light sources suitable for the internal lighting of buildings.

So, bearing in mind the cost factor, what do we have in the way of options that are suitable for public lighting?

4.2 LAMPS

For the purposes of street and communal area lighting, there are four main sources: sodium vapour, mercury, fluorescents and tungsten. Some explanation is necessary here in order to establish a basis for choosing the light sources that could be suitable in certain situations.

Sodium discharge lamps

Low-pressure sodium lamp

This type of lamp is recognised by most people as the lamp that produces a very yellow light. It has a very high efficacy which makes it popular with local authorities, but very poor colour rendition which makes it very unpopular with pedestrians, although motorists often prefer it. In fact it is a monochromatic yellow light which distorts virtually all normal colour characteristics.

High-pressure sodium lamp

The difference between the two types of sodium lamp is that here the sodium vapour is produced at a higher pressure. This makes a very considerable difference to the characteristic of the light produced and is far more acceptable in community lighting. Here the lighting is not monochromatic but pulls in colours from other parts of the spectrum and thus improves colour rendition greatly. The main drawback of high-pressure sodium lamps, as far as local authorities are concerned, are the increased running costs.

However, lamp technology is continuously making leaps forward, and the recent great improvement of high-pressure sodium lamps over low-pressure ones has led to a new generation of these lamps known as White Sons, which are now in common use. It is characteristic to refer to these types of lamp in terms of SOX or SON. These terms stem from their production but it should be noted that the term SOX refers to low-pressure sodium lamps and SON refers to high-pressure sodium lamps.

These two types of lamp are by far the most common in use for public lighting, particularly in Britain where low-pressure lamps were introduced some forty to fifty years ago. This situation is not true of all countries. America, for example, uses a considerable percentage of mercury lamps.

Plate 8 Street lit by low-pressure sodium lamps (SOX)

Plate 9 Street lit by high-pressure sodium lamps (SON)

Mercury lamps

Not long ago mercury lamps produced a distinct greenish hue which did little for people's appearance and therefore rendered the lamps unpopular. But much of that has changed and the modern mercury lamp, which operates on a similar principle to its counterpart, the sodium lamp, has greatly improved colour rendition. However, perhaps the most common form of using mercury now is to combine a high-pressure mercury discharge tube with tungsten filament. The filament performs as a ballast to the discharge tube but also increases its ability to produce colour at the red end of the spectrum, thus improving its colour rendition and making it an even more acceptable choice for community lighting.

The mercury lamp's efficacy and running costs are good. Even if the mercury lamp never becomes a serious competitor to the sodium lamps it has real value as a complement to both low-and high-pressure sodiums. Used sensitively it can create very considerable variety and interest when interposed with other types of lighting, such as sodium lamps and fluorescents.

Fluorescent lamps

In a fluorescent lamp the majority of light is produced by the discharge activating phosphors which are coated on the inside of the tube. The phosphors can be mixed to produce a range of colour renditions which give this type of lamp an edge over some of its rivals.

The use of fluorescent lamps in external public areas is limited by their size and vulnerability. However, given adequate protection and careful design, they can be used to good effect in enclosed communal areas such as corridors, entrance foyers and car parks where their good colour rendering can be of great use in face recognition. Compact versions of fluorescent lamps are now widely available and can be used to replace less efficient filament lamps in bulkhead lights and in fittings to illuminate entrances to individual dwellings.

As with mercury lamps, fluorescent lamps do not constitute a major element in community lighting in open areas but they have a very useful part to play both for the reasons given above and as a means of creating variety when used with other lamps. This can help to define certain areas or to give direction and clarification in such places as large housing schemes which are sometimes confusing, especially at night. Recent research by the Dutch company 'Industria' suggests that fluorescent lighting at comparatively low levels of illumination is more effective in street lighting for safety and recognition than it is given credit for.

Tungsten filament lamps

Although filament lights (where an electrical charge passing through a thin coil of metal generates white heat) are the most common lighting source for domestic use, their comparative inefficiency and vulnerability mean that they have limited application outdoors and in communal areas. However, they do give a 'warm' light with good colour rendition, and their easy availability for replacement means that they have their uses where good supervision is available (such as individual dwelling entrance porches).

Special applications

Areas such as sports and recreation spaces need much more intensive types of lighting. Tungsten/halogen lamps give an extremely bright light and can be used where 'daylight' conditions are needed. However, the intensity of the light can cause glare. They are expensive to run and are vulnerable, but a well-positioned tungsten/halogen lamp is unbeatable for highlighting.

Low-cost versions of this type of lamp are now widely available (e.g. through DIY stores) and, combined with a movement sensor, they are promoted as 'security lighting'. These are increasingly being installed by householders to illuminate back and front gardens, driveways or yards. Their sheer intensity and overspill means that they sometimes become supplementary private 'street' lights. This can, however, have a detrimental effect if the overspill causes a nuisance to neighbours. It can even be dangerous if the light is directed along a road and dazzles motorists. This can easily happen if a lamp is fitted to a dwelling situated on the corner of a road or at a road junction.

Lanterns

This general introduction to light sources suitable for safety lighting needs some explanation about their relationship to what they are housed in, namely the lantern or fitting. There are some lanterns which do not have in-built lighting reflectors, usually for specialist or decorative reasons, but the lighting output is usually very poor from such fittings. Modern fittings usually house reflectors which not only throw the light in a positive and directed way to give greater control over the light output, but greatly increase the efficacy of the lamp. Some sophisticated lantern designs incorporate prismatic refractors and lenses to direct and focus the throw of light.

4.3 CONTEXT AND PLACEMENT

However important lighting may be in promoting community safety, it is best used and understood as one element among others.

This multi-faceted approach can be difficult to implement, especially in existing run-down areas and where budgets are allocated for specific remedies. But a quick-fix approach often entails a waste of money or the value of the lighting being reduced. Probably the most obvious example of this is the lack of any attention to trees and bushes. It makes no sense to instal new or improved lighting if the light is going to be obscured by foliage. Where possible, the light should be placed somewhere else, the trees pruned back and maintained or, if necessary, removed and replanted elsewhere.

If starting from scratch with a new development, the problem is more easily dealt with. Here all elements of the built environment should be considered as an integrated whole. Planting should be of the kind that, when fully grown, will not interfere with lighting or with sight-lines through streets, alleyways, etc. Similarly, lighting points should not be placed in close proximity to property where the lighting could cause nuisance to residents in bedrooms at night. All design is a matter of compromising conflicting requirements to the best overall advantage.

What are the major elements to be considered for a cohesive environment? Lighting, as we have seen, is of great importance, particularly as a means of reducing fear and crime, but used imaginatively it can have great aesthetic benefits. It is not only monuments and important buildings that can be lit to produce interesting effects but light played onto the end terrace of the most humble dwelling can richly enhance the general surroundings. The integration of lighting with the layout of roads and pathways is crucial. Pretty winding pathways that look so charming on a drawing board are more difficult to light evenly and may produce 'blind spots' which make people feel more vulnerable.

The interplay of pedestrian and vehicular surfaces with grassed and cultivated areas is, of course, a main concern of the landscape planner, but again consideration should be given to these areas in relation to their night-time effect. In general there should be no shadowed hiding places, such as dense foliage or protruding walls, adjacent to footpaths.

Textures that can look so good during the day can visually disappear under certain lighting at night. The lighter, and therefore the more reflective the surface, the more more reassuring it is at night. This particular consideration is most important when lighting areas such as pedestrian tunnels and subways. The difference between the lighting on a light-coloured surface and a dark one can be the difference between quite adequate lighting and fearful darkness.

A further dimension to this particular problem is the choice of colour. As we see from other more detailed explanations in this book, colour rendering from communal lighting is, by the standards of daylight, very poor indeed. This has its most direct effect on the surface colour in any environment and it is not just a question of aesthetics. Signage systems that rely on colour coding, and which look splendid under daylight or on the designer's drawing board, can become totally useless under such lighting as that produced by sodium vapour lamps simply because the colours become completely changed. Decorative treatments like murals that sparkle and enliven a space during the day can look meaningless, even ugly, under adverse lighting conditions.

Lighting integrated with other items of street furniture

Under the general heading of street furniture comes a whole array of equipment from street signs to litter bins to seating, bollards, fencing, plant tubs and, of course, lighting columns.

Integrating all these elements is a task rarely undertaken successfully, if at all. Instead what we normally see is an incoherent approach which deals with parts of the overall problem as they arise. Consequently, most of our towns and cities are a visual mess of elements piled one on another as the money becomes available and the inclination takes the decision makers.

Seating

The relationship between public seating areas and lighting is particularly important when seen in the context of potential anti-social behaviour. True, people rarely sit on street seating to read at night. The real reason for establishing good lighting around seating is that seating areas are popular places for vandalism and seats are probably the most vandalised of street furniture. Link this to the fact that good lighting deters vandalism and you immediately reduce unwanted behaviour, costly damage to equipment and create a feeling of reassurance to those using the area for proper purposes.

Litter bins

Much the same reasoning should be applied to the siting of litter bins. Those using the seating and who do not have a litter bin in close proximity are often tempted just to drop litter around the seating area. We, therefore, are tempted to simply group the lighting, the seating and the litter bins together. Unfortunately there is one drawback to this. Litter bins are apt to smell and should not be located too close to seating. However, modern lamps are extremely efficient and can have a considerable throw of light around an area. Providing the bins are located fairly close to the seating and lighting, you can produce a catch-all compromise that will produce a successful solution with many social and economic advantages. The most obvious way of achieving this is to attach litter bins to lighting columns, which is both economical and avoids clutter.

Bollards

Essentially bollards, like many other elements within the built-environment, are about control, either of traffic or people. But their use does not stop there; they can be very effective as a source of lighting to define a perimeter at night. From a safety and preventive crime angle, lighting bollards placed within bushes have proved to be a great deterrent in addition to producing attractive effects with foliage at night. However, because bollards are at an accessible level for prying hands and flying boots, any lighting incorporated needs to be of a particularly secure and vandal-resistant design.

Bus stops

Lighting should be carefully positioned to illuminate the area around bus stops and shelters. Well thought-out lighting can not only illuminate waiting and alighting passengers but also timetables, information and advertisements.

Telephone boxes

Wherever possible it is advisable to light more than just the interior of telephone boxes. They attract vandalism otherwise, and people waiting to use the telephone could also benefit from some light.

Planters and plant tubs

On many housing estates planters are prone to misuse and vandalism. Not only are plants destroyed but the tubs are used for dumping rubbish. However, if ways can be found to prevent vandalism, by careful positioning and design, planters can do much to improve the visual amenity. Integrating lighting with planters provides a further opportunity to increase this amenity, and is likely to reduce abuse.

Essentially there are two ways in which this can be done. By integrating a normal column type of lighting into the planter which will throw general light around the area of the planter or, literally, to plant the lighting in among the foliage. The latter may help to alleviate the fear of predators 'lurking in the bushes' and will also help to add to the general pleasantness and sense of well-being of the area.

Walls and fencing

The placing of lighting with regard to fencing is most important. It is useful, where possible, to integrate the lighting with the fencing. This, at the very least, can reduce the clutter of objects within the area, but great care must be taken to avoid creating shadows, the offender's delight.

Co-ordinating the elements

One of the reasons why so much of the urban environment looks cluttered and incoherent concerns the organisation of public services and utilities. Each department or utility provider has its own particular responsibility, which tends to be independent of all other provision within an area. This leads to the classic 'gas board digging up the road two days after the highways department has resurfaced it' syndrome. In terms of lighting this can lead to total negation of improvements, such as trees growing up to obscure lamps or new walls casting footpaths into deep shadow. Co-ordination at the neighbourhood level is essential if the benefits of individual services are not to cancel each other out. This co-ordination can be extended beyond development work into maintenance – for example, the department responsible for lighting could issue light fault referral cards to other council employees who are out and about in neighbourhoods and estates.

4.4   PRIORITIES WITHIN AN OVERALL COMMUNITY SAFETY STRATEGY

Ideally we should not have to prioritise the value of the elements that go to make up a safer environment. But we do not live in an ideal world, so arguments have to be made within budgets and resources as to where the best value for such investment lies.

Over the last few years in the UK, and during the 1960s and 1970s in the USA, there have been numerous resident consultations about the impact of crime and fear within communities. Of those that have addressed themselves to the problem of the physical environment, it is very rare not to find residents putting communal lighting improvements within the top three of their priorities, and often at the top of any 'worry' list.

Again, when new lighting has been installed and upgraded within an area there is usually a large proportion of residents who believe that crime reductions have been effected and they feel safer about their environment. This leads to a greater confidence about going out at night in areas which, after dark, were formerly perceived as 'no-go'. The wider benefits of this increased activity should not be overlooked and can lead to the 'crowding out' of crime opportunities by the overwhelming presence of law-abiding citizens.

But what of the true costs of lighting as a priority within a community strategy? Compared to other major strategies – such as increased policing, reducing the height of tall buildings, taking out walk-ways, knocking down walls, building walls, removing trees and so on – lighting improvements are a comparatively economical option and, indeed, few of the major measures come anywhere near the low running costs incurred by lighting.

The only time lighting costs come close to most other crime preventative or remedial measure is in the installation of new column or post lighting where trenches have to be dug to install cables. But this rarely constitutes more than a small part of any new or upgraded lighting scheme. Very often only new lanterns or luminaires need to be installed and in many cases little more is needed than improved lamps or another few fittings where cable runs are already installed.

But the true saving in both monetary and human terms can be even more graphic when a bad situation can be rectified by the installation of one or two lamps. In such a situation, failure to install some lighting, leading to assault, means that the cost of hospitalising one person for just a week is around 250 times the cost of running a lamp for a year. Commercial lamps such as the type normally used in street lighting use far less electricity than domestic or interior type lamps.

However, costs are not the only considerations. Reliability, speed of effect, acceptance by users, useful life, etc., are all factors which should be taken into consideration when weighing up competing environmental design options.

On the issue of reliability, modern lamp technology has seen enormous improvements over the last few decades in all kinds of ways, but in few areas has lamp technology advanced more than in reliability. It is now rare, indeed, for a lamp to fail within its given lifespan, although this should not be taken by lighting authorities as an excuse for not having a regular monitoring system for light failures.

As to the speed of effect, those not used to being involved in lighting installations are often amazed at the speed with which lighting columns can be erected. Whole streets and areas can have new lighting columns in place within days and the installation rate is very much higher for such things as wall-mounted fittings. And, when finished, all that has to be done is to flick a switch and the effect is instantaneous.

The acceptance of improvements by the users has become a growing issue. There are many who object to the design of new building works for a variety of reasons, including costs, which they see reflected in their taxes. Or they object to being moved while their own homes are renovated as part of schemes being carried out. The main complaints against lighting improvements have come from people who, through bad lamp placing, have found increased lighting, usually into bedrooms, an unwanted intrusion. Such a problem is remedied easily. Apart from disapproval of lighting improvements by street crime specialists such as drug dealers and pick-pockets, objections may occasionally be raised by conservationists who are concerned about maintaining the appropriate design for additions to the street environment.

Rejection by people who have to live with design decisions can be acrimonious and costly when feelings run high. It is not always possible, but resident consultation on such matters can save money, time and effort in the long term. If residents are consulted before work is undertaken, there is usually goodwill towards any new scheme in which they feel they have taken some part.

Assessing the need to find the right solutions

The knowledge about what is needed to improve lighting in a given area does not rest solely with the lighting engineer. Traditionally, any installation of street or communal lighting has been planned by engineers according to objective standards of light intensity and coverage. However, as we have shown, lighting has a subjective, as well as an objective dimension.

Police, key workers and, particularly, residents have daily experience of the effect of artificial lighting in their neighbourhoods, and their views on shortcomings, vulnerable areas and possible solutions are most valuable. Given that lighting upgrades are usually undertaken on a limited budget, it is essential that the improvements made have the maximum beneficial effect. A careful reconciliation of engineers' objective recommendations with local views of what will work best is likely to produce the optimum cost benefit.

4.5   TECHNICAL CONSIDERATIONS

Artificial and natural light

Until comparatively recently, artificial light was a poor substitute for natural light. Most buildings in northern countries were designed to optimise the penetration and distribution of precious natural light. Even nowadays natural light is the preferred option in most settings, and this should be borne particularly in mind when designing or redesigning communal areas such as footpaths, passages, subways, entrance lobbies and corridors. Natural light is 100 per cent reliable (while the sun is up!) for these areas, and costs nothing to maintain. Why create darkness unnecessarily? However, modern artificial lighting is capable of revolutionising attitudes to the built environment for the better, both practically and aesthetically, and it can do it in two ways. Firstly, it can turn the use of lighting for buildings inside out: instead of relying on outside natural lighting to enter the building, the building can create its own lighting from within, often to affect the outside. Secondly, the outside of buildings can not only be made visible and attractive at night by the use of external lighting, but can dramatically help to produce environs in which people can move around more safely.

The distinction between interior and exterior building concerns are not always quite as straightforward as might appear at first. The main problem is the eye's slowness in adjusting to changing light intensities or coloration. This is most common when moving from a well-lit interior space at night into a comparatively gloomy external or communal space. For a while we may feel only 'partially sighted', which becomes a particular problem as we grow older. (A similar 'temporary blindness' can occur if we move from a sunny daylight exterior into a gloomy subway or multi-storey car park.)

True, many of the light sources used internally, such as tungsten and fluorescents, can be used externally, provided they are properly housed against weather, but rarely can light sources created for outside use be used internally. Basically this comes down to cost. The best external lighting, although constantly being improved, would be totally unacceptable in most interiors because of its poor colour rendering. However, the cost of running such light sources is often a fraction of normal internal lighting.

This transition from the interior of a building to its exterior (and vice versa) is of course a matter of concern to all good architects in terms of relative spaces and volumes and the choice of materials. But it should also be a matter of concern for those designing the lighting.

For example, a very serious problem exists around many railway, underground and bus stations. The problem usually arises because in the UK these areas are normally very well lit, whereas the immediate vicinity is lit to the lower standard of normal external lighting. This produces complaints from those using these areas; on leaving such brightly lit areas they are confronted by such relative darkness that they feel distinctly apprehensive, which, from a safety aspect, is quite justified. Where control is possible, the aim in situations such as these should be to gradually reduce the light as one moves away from the brightly lit area to allow some time for the eye to adjust. Indeed, this should be a consideration in all situations where there is a transition from high interior lighting levels to lower external levels. But the considerations would be much better if they were about quality as well as mere quantity. Here a simple mix of internal light sources such as tungsten with external sodium lamps could be most effective in helping to create a pleasing and aesthetic solution to this problem.

Crucial factors for community lighting

Sensible levels of illumination

Although the aesthetic aspect of community lighting should not be ignored – it has much to offer in terms of enhancing the quality of environments – the overriding concerns with community lighting in respect of safety will always be those of function and practicality.

The first of these practical concerns for most people will be: 'How much light is necessary to be effective against crime and to instill a feeling of safety?' The straight answer is, 'Never enough'. Street crimes of all kinds are committed at all hours of the night and day, and although these crimes might be less during the day, daylight – which is many times brighter than any artificial street lighting – does not completely stop such crime altogether.

However, studies that have been carried out show some quite dramatic crime reductions in areas where lighting levels have been significantly raised, and almost all users say they feel better in such areas. The decision on the amount of lighting usually involves a balancing act between what can be effective and what can be afforded. Belgium, for example – probably the best lit of all countries – takes the view that a larger percentage of resources should be spent on lighting than in other countries.

In the first strictly controlled lighting and crime evaluation carried out in Europe, undertaken in Edmonton, north London, the lighting levels were established by using the lux unit of measurement to be to read at 1.5 metres above pavement level. The mean lighting level was established at an average of 10 lux with an attempted minimum of 5 lux.

The establishment of these two co-ordinating readings were based on several premises. The project, unlike all the American projects before it, was designed to be as small as could be creditable but, at the same time, as scientifically controlled as possible in a public street. The project was to concern itself primarily with the effect of lighting on crime reduction as a statistical end in itself. This, of course, demanded that all measurable aspects of the project should be as strictly predetermined as possible.

Therefore the two crucial measurements, the light level and the position of the reading, became paramount.

The level, to be of practical use, had to be of an appreciable level above what had already existed in the test area (a normal side street level of around 2–3 lux), but not so high as to be unrealistic in terms of running costs – thus the establishment of the levels of 5 lux minimum and 10 lux as an average. The measurement height of 1.5 metres is the average height of the human face. This element is crucial, as facial recognition is the essential element in the effect of street lighting on behaviour.

The establishment of these determinants had a very considerable effect on crime reduction within the test area. Since then a considerable amount of work has been carried out in the UK, particularly in the London area, using higher lux levels. Most of this work has not been carried out with evaluations but has clearly shown a growing desire among most residents for greatly increased lighting levels.

Plate 10   Sensible lighting to balcony flats – Vauxhall, South London

Of the 115 lighting and crime evaluation projects carried out in the USA in the 1960s and 1970s, it is important to say here, while discussing lighting levels, that lux levels played an important part in all the experiments. However, because the work is so wide-ranging and is often done with different aims in mind, it has become quite complicated to extrapolate the light levels factor from all other factors.

However, based on these American studies, it has been estimated that 42 per cent of all night-time street crime takes place at lighting levels of 5 lux or less; another 30 per cent takes place between 5 lux and 10 lux; and so on until you get to very high levels of around 20 lux at which only some 3 per cent of night-time street crime takes place. Whatever is made of these figures, the correlation between the figures for crime and levels of lighting simply cannot be ignored.

To those who ask 'What level of lighting should be used?', the biggest pay-off is at the bottom end of lighting levels. Raising the lux level from an average low level of 2–3 lux by 6–7 lux makes a dramatic difference, but the benefits do seem to tail off in terms of crime statistics, as the levels get higher. But, referring to earlier statements about evenness of illumination, it must not be forgotten that much of the benefit from increased lighting would be lost if the levels achieved were not consistent throughout the area. Indeed, there is some evidence that crime is more common in highly but patchily lit areas than in comparable areas with lower but more evenly distributed lighting.

Colour rendition

As we saw earlier, the most basic form of lighting is monochromatic, where all visual experiences are seen by reflecting one colour source. Through evolution our eye has come to regard the best source of light as daylight. Here most of the colours in the spectrum are reflected back from the object to our eye, thereby giving a 'complete' picture.

The reason that this is so important in crime prevention and security is primarily *recognition*, particularly of faces which display very subtle tonal and colour nuances. Recognising, and being able to recall what has been recognised, is a highly complex human ability and does not simply rely on the amount of light around. It also relies on the quality of that light, and by quality we mostly mean colour rendition.

It would be easy to say that the difference between good and poor colour rendering is the difference between a black and white photograph and a colour one. But this would be a wrong comparison because in black and white photographs there is no colour confusion at all, nothing of a colour nature to distort our recognition of someone we know. However, even in the poorest lighting there is always some colour, and while our eye is able to instinctively make adjustments and allowances for the black and white photograph it is very confusing to see someone we know whose colouring has been changed by the quality of the light.

Our abilities to see and compute what we see is nothing short of astonishing. When we see something like a head or face we take in masses of information. The size of the head. The shape of it. The textures involved from the skin type to hair. The arrangement of features such as large eyes or a small nose, and so on. But chief among the features we recognise, that makes a person unique, is colour. Even those who would not normally regard themselves as having good colour sense, take in tremendous amounts of colour information about the appearance of an individual.

The poorer the lighting quality (the poorer the colour rendition of street lighting), the less able we are to recognise and discriminate. This produces fear and uncertainty in those going about their business lawfully but confronted by those they feel wary of. And those with criminal intentions become instinctively aware that poor lighting affords protection against being recognised.

Evenness of illumination

The third crucial factor in community lighting and safety, alongside the level of illumination and colour rendition, is the evenness of a lighted area. These three factors are inseparably intertwined and all good lighting schemes must take account of all three. However, evenness of illumination is by far the most easily understood, in that the aim is simply to light an area to the same consistent level in all parts.

This, of course, is not totally possible as light levels decrease the further you move away from any light source, but it is possible, by good planning, to minimise this problem, and by the use of computer plotting, of which most lighting companies are capable, it is possible to forecast very accurately the levels and evenness of any lighting scheme.

Planning

So, with these three crucial factors in mind how should a good community lighting scheme be planned? One of the real problems with which community lighting (or 'lighting for people') has to come to terms is that, generally, lighting schemes have to fit into existing built environments. The joy of planning a suitable lighting scheme along with planners and architects at the inception stage is something rare to those concerned with lighting design. Neither for them is the relative simplicity of wide open uncomplicated areas such as tennis courts and motorways.

Lighting public areas in communal and housing environments is full of the complications of undulating façades, narrow alleyways, courtyards, archways and indeed all the building forms that go to make up dense living areas. So this does not just require the knowledge of what is technically desirable, such as lighting levels, but also requires considerable sensitivity in such things as where lighting is placed. Indeed, if there were to be a fourth crucial factor it would be the placing of luminaries in the environment. The problem here is that rules cannot be universal in the way level, colour rendition and evenness are universal. The placing of luminaries will differ from one situation to another and relies largely on the sense and sensitivity of the lighting designer or engineer.

But even with such a complex problem there are guidelines and recommendations that can be applied to most situations.

Firstly, the task is made much more simple if the measurements planned for community lighting are carried out in a way that is different from the method of practice of most lighting designers. Bearing in mind what was said earlier about the levels of illumination being measured at 1.5 metres from the ground (i.e. at face level), then for community lighting, levels should be designed to function at this point. The positioning and output of the luminaries can then be extrapolated.

Glaring errors

An old lighting engineer's adage is that 'one person's sparkle is another one's glare'. There is in most sorts of lighting a factor generally referred to as 'sparkle'. This is purely a qualitative phenomenon and most difficult to verbalise. But we all recognise it when we see it. It is that quality we attach to cut-glass when it seems to flash small elements of light. Without it, especially in interior settings, lighting looks flat and depressing.

However, as in most phases of life, you can have too much of a good thing. If sparkle is taken to excess it starts to produce another qualitative attribute known as 'glare'. Glare is much easier to define, as we not only all experience it but have suffered from it – it is blinding and often quite hurtful. In interiors it is a nuisance, perhaps worse than no lighting at all. In lighting for community safety it can be both a nuisance and dangerous.

Dazzle blindness, however temporary, in an unsafe place will produce feelings of fear and anxiety, and although it may be difficult to eliminate glare completely it should be reduced to the minimum without getting rid of sparkle altogether.

But how should this be done? Fortunately today there are many luminaries and lanterns on the market that deal with this problem admirably and every effort should be made to use such equipment. The problem mainly arises when powerful point light sources are being used. It is often just the sheer power of the light that causes the glare. So, as a general principle, try to keep such light sources from being installed very close to where people might use the area and try at all costs to place such lighting well above eye level. This eliminates glare by not letting the light directly enter the retina.

If the problem of glare is the greatest, it is by no means the only one. Far less common, but still a very great nuisance when encountered, is the use or perhaps the misuse of what is generally termed 'floodlighting'.

There is a naive belief that if you erect enormously powerful floodlights on the tops of tall buildings and try to simulate the sun you can somehow scorch the criminal element out of an area. There may be evidence to say that this should work, but the cost to the vast majority of law-abiding people living in the area is too great to be acceptable.

Where tall buildings used for such purposes are in close proximity to other buildings, the problem created by lighting spilling from the floods on to these other buildings, and therefore into people's bedrooms all night, is something that must be avoided. It is one thing having people sleeping uneasily through fear of crime in their area, but quite another if they cannot sleep because of the community lighting.

Usually such relatively crass solutions produce other problems in that such strong illumination from one direction produces strong shadows and will tend to light the tops of people's heads rather than their faces. However, it would be quite wrong to damn the use of floodlighting completely as a means of reducing crime and fear as some very successful schemes have been carried out using such lighting. These, in almost all cases however, have been where tall buildings stand quite isolated from other buildings and can light up a wide area around the building without creating nuisance. But it is rare that such lighting can be relied on solely and without the support of other lighting to eliminate shadow spots.

CHAPTER 5

ENTRANCE DOORS AND WINDOWS

INTRODUCTION

This chapter is concerned with upgrading and improving the security of entrance doors and windows. It attempts to provide a comprehensive guide to entrance door and window security by highlighting the potential risks and then by offering practical guidance to reduce and overcome security problems. The guidance is presented in a 'pick and mix' format designed to enable packages of security improvements to be drawn up to suit particular situations and potential risk.

It is important at this stage to highlight a potential conflict between the security of buildings and the needs of fire safety – in particular, the means of escape in the event of a fire. Of course, fire safety must take precedence and even more so in certain types of building such as multi-storey blocks of flats where an individual entrance door to a flat is usually the only means of leaving the dwelling. It is also important, however, that a reasonable level of security is provided so that residents do not feel the need to fit additional security measures. Measures such as locks that are key operated from both sides of the door may hinder escape and hamper access by the fire service. Furthermore, if these locks are fitted to fire doors, they may easily undermine the resistance of the door itself.

Section 5.6 of this chapter describes door locks for fire escape purposes. However, it is strongly recommended that the local fire officer and the building control officer be consulted at an early stage when proposing to carry out any security improvements. After all, it has been known for windows to be included in fire escape routes.

ENTRANCE DOOR SECURITY

5.1 VULNERABILITY OF ENTRANCE DOORS

The vulnerability of an entrance door will depend on a number of factors. These include:

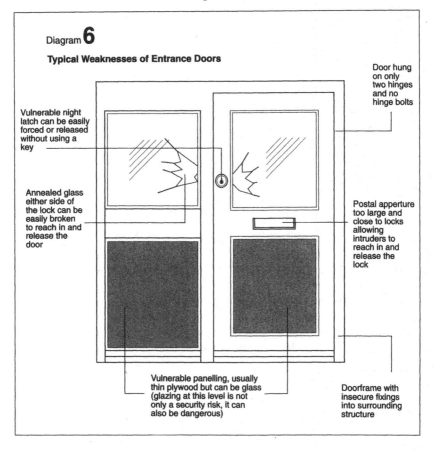

Diagram **6**

Typical Weaknesses of Entrance Doors

Door hung on only two hinges and no hinge bolts

Vulnerable night latch can be easily forced or released without using a key

Annealed glass either side of the lock can be easily broken to reach in and release the door

Postal apperture too large and close to locks allowing intruders to reach in and release the lock

Vulnerable panelling, usually thin plywood but can be glass (glazing at this level is not only a security risk, it can also be dangerous)

Doorframe with insecure fixings into surrounding structure

- *Location of door* Those situated in secluded positions, such as the rear or side of a dwelling, tend to be more vulnerable than entrance doors which are clearly visible and overlooked by neighbours, pedestrians or passing motorists. The levels of external lighting, the nature of the surrounding landscaping and the suitability of fencing, hedging or walling are also important considerations when assessing entrance door vulnerability.
- *Type of door and its components* Entrance doors can be subjected to many types of attack, including manipulation of the locking devices and impact force. Diagram 6 illustrates some of the most common weaknesses associated with entrance doors. The level of security provided by an entrance door will therefore depend upon its ability to resist any anticipated attack.
- *How security devices are used* Experience suggests that clear operating instructions, and even demonstrations, may need to be given to residents on the use of entrance door security. Possibly one of the most common problems of security improvement schemes is that extra locks are fitted to doors but not used.

The guidance this book offers for security improvements is based on three principles:

- External access and manipulation of locking devices must be prevented.
- All components of the entrance door (and frame) must hold together if forced entry is attempted.
- All security devices must be easy to operate with clear instructions given in order to encourage use.

5.2   THE ENTRANCE DOOR

There are many different types of entrance door available. Timber doors, both softwood and hardwood, still account for the majority of doors in use today. This is not surprising given timber's versatility. It is also a very adaptable material, which means that it is relatively easy to fit additional security locks to timber doors.

There are, however, other materials from which doors can be made. For example, steel framed doors were used typically as 'French doors', but these have generally fallen out of favour because they were found to be draughty and can be difficult to maintain. They have also been superseded by the rise in popularity of the aluminium door or, more specifically, the aluminium patio door. Aluminium has the advantage of being lighter in weight, but, more importantly, requires little in the way of maintenance such as painting. uPVC doors are becoming increasingly popular. They have similar virtues to aluminium in that they require little maintenance, but they also have better thermal properties and are available in a wide range of designs. In addition, uPVC doors usually feature multiple locking devices which are built into the door (during manufacture) and which, it is claimed, offer a higher degree of door security.

Entrance door designs

Although there are many types and designs of entrance door, for ease of presentation they have been divided into a number of groups, as shown in Diagram 7.

Hollow flush doors

These usually consist of thin plywood facings fixed to a flimsy timber frame with cardboard box infill. They are available with or without glazed panels. These doors are insubstantial, providing little resistance to force or opportunity to fit additional security hardware. These doors should not be specified for external entrance doors to any building and existing ones should be replaced.

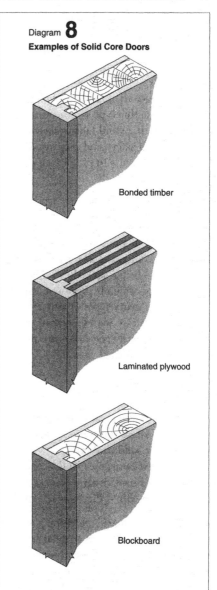

Diagram **8**

Examples of Solid Core Doors

Bonded timber

Laminated plywood

Blockboard

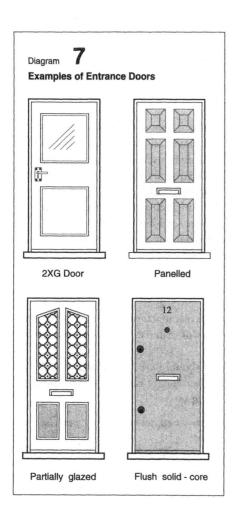

Diagram **7**

Examples of Entrance Doors

2XG Door

Panelled

12

Partially glazed

Flush solid - core

Fully glazed doors

These usually consist of an outer frame of timber (or steel, aluminium or uPVC) supporting a large glass panel. The common 2XG door, which is similar to the door above but has a central horizontal rail, has also been included in this category. One of the main attractions of this type of door is the amount of natural light they allow into a dwelling. They are also relatively inexpensive, depending on the type of glazing used.

The level of security they provide is debatable. Larger panes of ordinary annealed glass in doors such as these tend to be less vulnerable than doors with small panels of glass on the grounds of the amount of noise created when broken. However, once the glass is broken it is relatively easy to gain access into a dwelling. Furthermore, this type of door can be dangerous if the glass is broken accidentally; for example, if a child falls against it.

In terms of safety and security these doors should be glazed with laminated glass, which will continue to provide an effective barrier even if it is broken.

Partially glazed doors

Probably the most common type of a partially glazed door is the 2XG door. This door consists of a outer frame with a central horizontal rail. The upper panel of the door is glazed and the lower panel is infilled with thin plywood which is particularly vulnerable to being 'kicked in'. These doors are frequently used as side or rear doors for dwellings, which makes them especially prone to attack.

There are many types of partially glazed door, including the decorative hardwood varieties, i.e. those with Georgian style windows or semi-circular glass panels. Structurally, hardwood doors are usually far superior but, again, they should be glazed with laminated glass.

Solid-core flush doors

'Solid core' is a general term for doors which are manufactured without any internal void areas. These doors are infilled with a variety of materials such as plasterboard, chipboard or solid timber.

For security purposes, solid-core flush doors should have been designed and manufactured to withstand impact and prevent bursting inwards. Doors of this kind shown in Diagram 8 can be made in the following ways:

- bonded timber strips forming a rigid raft, sandwiched between 9 mm exterior grade plywood facings and hardwood lippings;
- laminated exterior grade plywood forming a rigid raft, with hardwood lippings;
- door stiles of at least 75 mm wide glued to blockboard, with 9 mm exterior grade plywood facings and hardwood lippings.

Solid-core flush doors are usually faced with plywood, although they are available with facings of other materials such as decoratively coated steel which has been designed to provide more resistance to abuse.

Fire doors

These can look similar to solid-core flush doors but have been designed to prevent the spread of fire for a given length of time. In order to be classified as a fire door, the door must have been tested by an accredited testing laboratory in accordance with British Standard BS 476. Fire doors are used extensively as entrance doors to flats in multi-storey blocks. For security purposes, they should also have been designed to withstand impact.

5.3 THE DOOR FRAME

There is rarely a need to replace door frames for security reasons unless they are damaged beyond repair. However, it is often necessary to improve door frames in order to up-grade security.

Improving door frame fixing

All external door frames should be fixed securely to the surrounding structure. Strong fixings are particularly important at the points where hinges and keep plates for locks are positioned. These are the areas of potential weakness if the door is subjected to impact force. A simple yet effective way to do this is with frame fixings which have been specifically designed for this type of application. These consist of a long plastic plug and steel screw which is inserted into a single hole drilled through the frame and into the surrounding structure. It is advisable to use at least four frame fixings on each side of the door frame.

Frame fixings can also be used to improve the fixings of door ironmongery such as hinges and keep plates. Diagram 13 on page 54 shows where an ordinary wood screw in a hinge has been replaced with a frame fixing which has been fixed through the frame into the surrounding structure. This approach is particularly useful on narrow door frames of 50 mm deep or less because these frames rarely provide an adequate depth of timber to which hinges, locks, etc., can be securely fixed. Ideally these should be replaced with more substantial frames, but if for any reason this is not possible, then all of the ironmongery for a door which is attached to the door frame should be fixed as described above.

Door stops

The depth of door stop for inward-opening doors should a minimum of 18 mm deep. Although this is now standard practice, many older door frames have door stops of 12 mm or even less. An 18 mm door stop provides more resistance to leverage and against attempts to push back night latches by forcing steel tape or inserting a credit between the door and frame. Door stops of less than 18 mm should be increased by fixing additional timber to the frame. If the existing door stop is cut from the solid wood, an additional strip of timber should be added to the frame. Alternatively, if the door stop is planted onto the frame, it should be removed and replaced with an appropriate timber strip (18 mm deep). In both cases, the additional timber should be glued and screwed in place, filled and painted over.

Door stops of 25 mm are usually specified for fire-resisting doors, and as they provide even more resistance they should therefore be used on particularly vulnerable doors.

Combination frames

Many entrance doors are incorporated into larger combination frames, which are likely to be glazed and have opening windows. They may also have vulnerable thin timber panels. All of these features can present a security risk, particularly if they are located close to the locking edge of the door (see Diagram 6 on page 45). To reduce these problems:

1. All glazing within the door and frame should be laminated glass. This glazing should be fixed in such away as to prevent it from being easily removed from the frame (see pages 94 and 95).
2. The locks specified for these doors should be key operated from either side of the door, although this will obviously depend on whether emergency escape requirements apply (see pages 66 and 67).
3. Thin timber panels should be replaced or reinforced with stronger materials such as 12 mm (minimum) exterior grade plywood, fixed in place with wood glue and screws.
4. Where new doors and frames are to be installed, the door should be hung on the side of the frame adjacent to the glazing and any timber panelling. This will make it more difficult to reach the locks through broken glass or panelling in the frame. Obviously, this will make little difference if the combination frame has glazing or panelling on both sides of the door.

5.4   THE HANGING EDGE OF THE DOOR

When entrance door security is being considered, emphasis is usually placed on improving the locking edge of the door. However, equal attention needs to be given to the hanging side of the door.

Conventional hinges

It has been standard practice to hang all external entrance doors to dwellings on three 100 mm hinges. However, this may not be enough to prevent the major failing of most conventional hinges if they are subjected to impact force or leverage. Diagram 9 shows how the narrowness of the hinge flap does not allow the screw holes to be staggered enough to prevent the timber splitting along the line of the screw. In fact, this weakness may already be apparent on newly hung doors, if care is not taken when fixing the screws.

   This is a common fault with the majority of hinges used on entrance doors – even where broad butt hinges are used (the screw holes are still positioned virtually in a straight line). Furthermore, the hinge itself may even fracture along this same

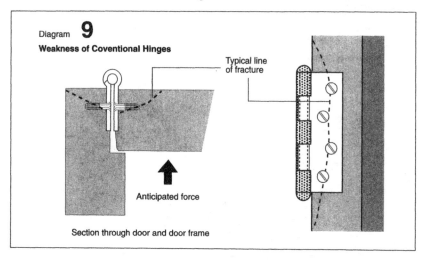

Diagram **9**

Weakness of Coventional Hinges

Typical line
of fracture

Anticipated force

Section through door and door frame

line if it is made of a brittle material such as cast iron. In these circumstances, hinges should be replaced.

Most entrance doors to dwellings open inwards. On less common outward–opening doors, the hinge knuckles are located on the outside of the dwelling and are therefore more vulnerable to attack such as leverage. The exposed knuckles could also be sawn off (although in reality the risk of this is very small).

Improving conventional hinges

On doors already fitted with conventional hinges, the risk of timber splitting on either the door or door frame can be reduced by the following methods.

- *Fitting hinge bolts* – probably the most common method. Hinge bolts, as shown in Diagram 10, protrude from the edge of the door and engage into keep plates fixed into the door frame when the door is shut. A minimum of two hinge bolts should be fitted to all external entrance doors. In most cases these should be fixed 50 mm above the bottom and middle hinges. On outward–opening doors a third hinge bolt should be fixed 50 mm below the top hinge.
- *Fitting hinge hooks* – which is a variation of the hinge bolt. Fixed onto the edge of the door, it is designed, as the name suggests, to hook onto a steel bar fixed in the door frame. This is shown in Diagram 10. The advantage of hinge hooks is that they require little timber removal to fit and allow for strong fixings into the door and door frame.
- *Fitting reinforcement strips* – which is an alternative and rather more expensive method of improving existing hinges. These act in a similar role to the reinforcement strips used to protect mortice locks. They clamp together the timber surrounding the hinge on both the door and door frame (see Diagram 11). The level of protection they provide is high, as long as good-quality hinges, secure fixings and robust doors and door frames have been used.

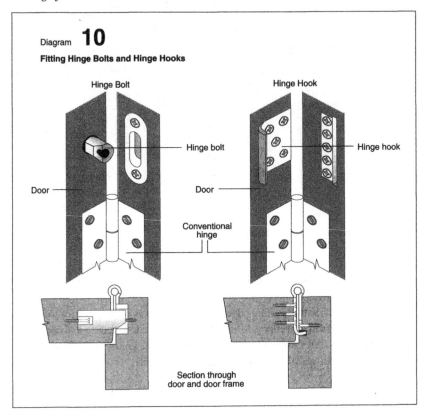

Diagram **10**

Fitting Hinge Bolts and Hinge Hooks

Hinge Bolt

Hinge Hook

Hinge bolt

Hinge hook

Door

Door

Conventional hinge

Section through door and door frame

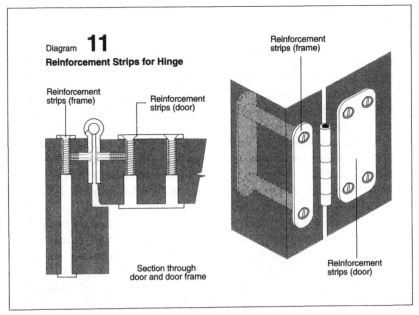

Diagram **11**

Reinforcement Strips for Hinge

Reinforcement strips (frame)

Reinforcement strips (frame)

Reinforcement strips (door)

Section through door and door frame

Reinforcement strips (door)

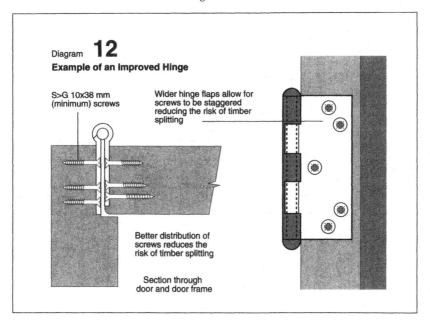

Diagram **12**

Example of an Improved Hinge

S>G 10x38 mm
(minimum) screws

Wider hinge flaps allow for
screws to be staggered
reducing the risk of timber
splitting

Better distribution of
screws reduces the
risk of timber splitting

Section through
door and door frame

Improved hinges

Where new doors and door frames are to be installed, there are hinges available
which have been specifically designed to improve the security of the hanging edge
of a door. These can be fixed to new doors and, in most cases, where a third hinge
is being added to an existing door (except where rising butts are to be retained).

1. Diagram 12 shows an example of a hinge which has been designed with a wider
   flap. This allows for the screw holes to be staggered in order to reduce the risk
   of the timber splitting. The wider flap also makes provision for five screws
   rather than the usual four. These hinges should be fixed using SG 10×38 mm
   twin-threaded screws or even longer where door and door frame allows.
2. Diagram 13 shows another hinge variation which has a number of features.
   This hinge has eight fixing screws per hinge flap instead of four found on a
   conventional hinge. Grooves also need to be cut in both the door and door frame
   to accommodate the hinge which is designed to reduce the risk of the hinge being
   ripped from the frame. Finally, the hinge incorporates a stud which functions in
   a similar manner to a hinge bolt. This stud can be fixed through the door frame
   into the surrounding structure.

Piano-type hinge

This hinge extends along the full length of the door and door frame and is fixed
through the door frame onto the surrounding superstructure with bolts. Although
rather expensive, it has been designed to provide high resistance to abuse from a

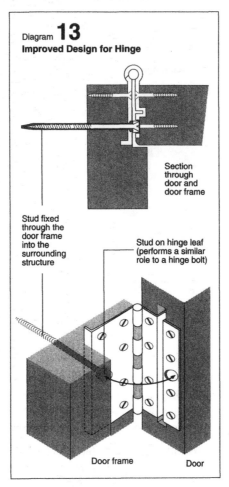

Diagram **13**
Improved Design for Hinge

Section through door and door frame

Stud fixed through the door frame into the surrounding structure

Stud on hinge leaf (performs a similar role to a hinge bolt)

Door frame

Door

determined intruder. Piano-type hinges are typically used on communal entrance doors to blocks of flats.

5.5 ENTRANCE DOOR LOCKS

Over recent years there have been many advances in the design and technology of locking devices for entrance doors. For example, a number of new locking mechanisms have been developed giving a greater number of key variations. With the advent of uPVC doors and frames, specific locking mechanisms have been designed which are 'built-in' to the door during its manufacture. Versions of these are also available to be fitted to timber doors. Some manufacturers have even dispensed with the conventional style of key altogether in favour of electronic systems involving the use of 'smart' cards or non-contact 'fobs' combined with magnetic locks. The application of this kind of technology has so far tended to concentrate on communal entrances to buildings such as blocks of flats rather than individual dwellings (see sections 6.4 and 6.5). As the cost of technology reduces, it is possible that this approach may become a viable option for entrance doors to dwellings. After all, smart card locks are already used extensively by hotels.

For the most part, and the foreseeable future, external entrance door security for dwellings will still rely on the use of conventional-type locks. These can be categorised in various ways. For ease of presentation, door locks have been divided into two main types:

1. Locks that are fitted onto the inner surface of the door, the most common being a cylinder rim lock. These usually use a pin or disc tumbler mechanism.
2. Locks that are concealed within the structure of the door, generally known as mortice locks. These usually operate with a lever mechanism although some are available with pin tumblers.

Cylinder rim locks

Possibly the simplest and least expensive version of this lock is the night latch which opens from the outside with a key and from the inside by tuning a knob. These locks are typically found on front or final exit doors from dwellings. They are made up of two parts, a lock case which is attached to the door and a staple which is attached to the door frame. There is a bevelled spring latch which protrudes from the lock case and engages into the staple in order to lock the door. By turning the key or knob, the spring latch is drawn back into the lock case to enable the door to open. When the key or knob is released, the latch automatically springs back into the 'locked' position.

These locks are vulnerable if the internal knob can be reached from outside – for example, by reaching in through postal apertures or broken glazing. They are also susceptible to manipulation by a credit card or steel tape. These types of cylinder rim lock should only be retained on entrance doors if an additional key-operated dead lock is installed (see page 57), measures are taken to prevent external access to the inner knob (see page 50) and improvements are made to the staple (see next page).

To overcome many of the problems of external access, locks have been developed which allow for the latch to be deadlocked. Variations of this include:

1. Locks which require the key to be inserted into the lock from outside the dwelling and then turned (one full turn of the key) in the opposite direction to that required to release the lock. This will then deadlock the latch ensuring that it can only be opened with the key.
2. Locks which have an auxiliary bolt (see Diagram 14). This automatically deadlocks the latch when the door is closed.

Most cylinder rim locks are available in two sizes (distance between the edge of the lock case and the centre of the barrel which contains the pin tumbler), *standard size* for use on the majority of doors and *narrow size* for use on doors which have stiles of less than 70 mm, such as 2GX doors.

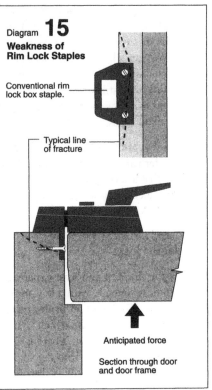

Diagram **15**
**Weakness of
Rim Lock Staples**

Conventional rim
lock box staple.

Typical line
of fracture

Anticipated force

Section through door
and door frame

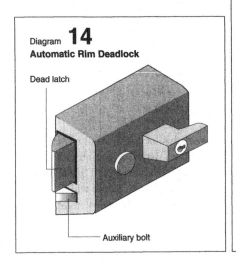

Diagram **14**
Automatic Rim Deadlock

Dead latch

Auxiliary bolt

Improving the performance of cylinder rim locks

One of the major weakness of locks fitted onto the surface of doors and frames is the ease with which the staples can be ripped from the frame if forced is applied to the door. This is illustrated in Diagram 15. The staples can be marginally improved by fixing them to the frame with longer screws, but even if the staple remains in position there is the risk that it could itself brake as they are usually made of brittle materials. It is often necessary, therefore, to replace the staple or fit additional hardware for reinforcement.

Replacing the staple   Diagram 16 shows a proprietary brand staple which has been specifically designed to overcome the problems listed above. This staple is made of plated steel and has angled lugs which are recessed into the door frame. These allow the staple to be fixed with screws, up to 44 mm in length, which are able to penetrate the full width of the door frame. This staple also has a better distribution of screw fixings, making it more difficult to rip from the frame. Another feature of this staple is the raised lugs which are designed to prevent manipulation of the spring latch with a credit card or steel tape.

Reinforcing the staple   Existing rim lock staples can be retained and improved by fitting reinforcement strips. Diagram 17 shows an example which involves fitting

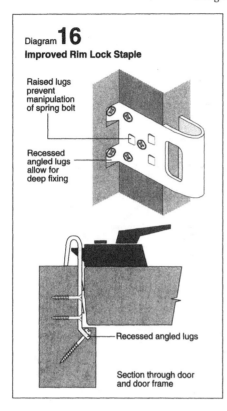

Diagram **16**
Improved Rim Lock Staple

Raised lugs prevent manipulation of spring bolt

Recessed angled lugs allow for deep fixing

Recessed angled lugs

Section through door and door frame

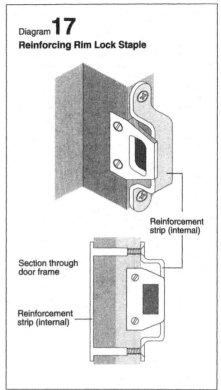

Diagram **17**
Reinforcing Rim Lock Staple

Reinforcement strip (internal)

Section through door frame

Reinforcement strip (internal)

two steel strips: one placed over the staple and the second on the outer surface of the door frame. Both strips are then bolted together through the door frame. These strips help to prevent the staple being ripped from the frame as well as preventing the staple itself from breaking.

Mortice locks

An alternative method of fitting locks to doors is to conceal them within the structure of the door. The most popular of these are known as mortice locks (see Diagram 18) of which there are two common types.

Mortice deadlock This has a square-ended bolt (dead bolt) which moves in and out of the lock case at the turn of a key. Mortice deadlocks are used typically as secondary locks on doors to supplement, for example, a rim lock. Some mortice locks have a double throw facility for use on double doors, especially those which do not have a rebated meeting (locking) edge.

Mortice sashlock This has both a dead bolt (as above) and a spring latch (similar to a night latch). It is typically used for rear and side doors. The spring latch is operated both sides of the door by lever handles and its purpose is to keep the door

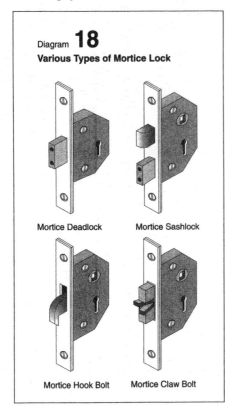

Diagram **18**
Various Types of Mortice Lock

Mortice Deadlock   Mortice Sashlock

Mortice Hook Bolt   Mortice Claw Bolt

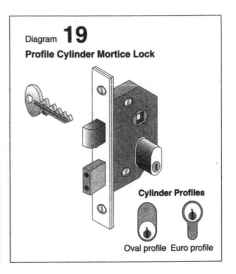

Diagram **19**
Profile Cylinder Mortice Lock

Cylinder Profiles

Oval profile Euro profile

shut. The dead bolt is key operated (again from both sides) and used to lock the door. A variation on this lock is where a spring roller bolt is used instead of the spring latch. The benefit of this is to make it easier to shut the door – the roller bolt is easier to engage than the latch.

Variations of mortice locks

There are a number of variations of mortice locks available (see Diagram 18) which have been designed for specific purposes.

Hook bolt and claw bolt   The hook bolt and claw bolt are two examples of locks which have been designed for use on sliding doors. The hook bolt version, as its name suggests, hooks onto the keep plate; whereas the claw bolt expands into the keep plate. To operate, both versions lock automatically (snap lock) as the door is shut and require a key to open.

Cylinder mortice locks   Mortice locks are usually made with lever mechanisms. Although initially more expensive, cylinder mortice locks have two main advantages over lever locks. Firstly, if keys are lost or stolen only the cylinder needs replacing.

This is particularly useful if there is a high turnover of occupants. Secondly, cylinder mechanisms provide a far greater number of key combinations (differs).

These locks are generally available with two different cylinder profiles (outline shape of cylinder), as shown in Diagram 19. Other cylinder profiles are available but are less common and are usually unique to a particular manufacturer. Profile locks can be key operated from both sides of the door. They are also available with a thumb turn on the inside (for the purpose described on page 63).

Other variations include locks with particularly narrow lock cases. These are used on aluminium or steel doors to shop and office entrance doors.

Locks made to British Standard

Locks bearing the kite mark are manufactured and submitted to the British Standard Institution for testing and approval to BS 3621: 1980: *Specification for thief resistant locks*. To meet this standard, locks must undergo a number of tests and have certain requirements, including:

- The lock must have a minimum of 1,000 key combinations.
- The lock must be resistant to operation by keys that are almost the same as the correct key.
- The lock must withstand pressure exerted against the end of the bolt to a load of 250 lbs.
- The bolt is shot (thrown) and withdrawn by the key 60,000 times at a rate of between 50 and 70 times a minute.
- The bolt must withstand the use of a pad saw for a minimum of 5 minutes.
- The lock is examined by a panel of experts for general vulnerability.

Virtually all guidance on security issued by the Home Office and individual police forces recommends that locks manufactured to BS 3621 should be fitted to external entrance doors. There are, however, many less expensive 5-lever mortice security locks on the market and it could be argued that these would perform just as well. Nevertheless, insurance companies are increasingly stipulating the nature of the security measures policy holders should have. Some even go as far as offering incentives such as reduced insurance premiums, or penalties such as reduced insurance cover. This approach is borne out in the guidelines issued by the Association of British Insurers (ABI) which recommends, for example, that only locks manufactured to BS 3621 should be fitted to entrance doors to dwellings, offices, etc.

Improving the performance of mortice locks

The removal of timber in order to install mortice locks on wooden doors invariably weakens the stile of the door and the door frame. This can lead to the timber splitting at the points where strength is needed. When force or leverage is exerted on the door, either the timber surrounding the lock splits away allowing the lock to fall out or the edge of the door frame splits forcing out the keep plate. Furthermore, using box keep-plates on inward opening doors can aggravate this problem by the

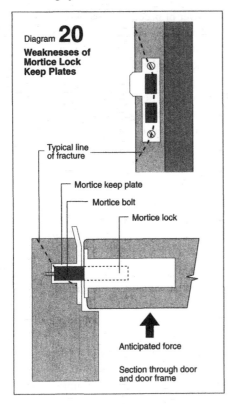

Diagram **20**

Weaknesses of Mortice Lock Keep Plates

Typical line of fracture

Mortice keep plate

Mortice bolt

Mortice lock

Anticipated force

Section through door and door frame

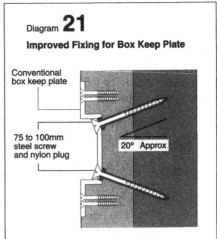

Diagram **21**

Improved Fixing for Box Keep Plate

Conventional box keep plate

75 to 100mm steel screw and nylon plug

20° Approx

need to remove timber to take account of the size of the box. Box keep plates are only really necessary on less common outward–opening doors, although they are usually supplied with British Standard locks. These weaknesses are shown in Diagram 20.

The performance of mortice locks can be improved as follows:

Improving the keep plate   Where a box keep plate is already in position, additional fixings should be introduced through the back of the box keep and frame into the surrounding structure. A single fixing should be used for the box keep plates of mortice deadlocks (single bolt) whereas two fixings can be used through the larger boxes of mortice sashlocks (a bolt and a latch). The latter should be fixed in a dovetail fashion as shown in Diagram 21.

To provide greater resistance to force, an angle iron can be fitted on the internal face of the door frame to reinforce the keep plate. The angle iron should be at least 400 mm in length, extending equal distances above and below the keep plate. The angle iron should be fixed using SG 10×45 mm (minimum length) twin-threaded screws at 100 mm centres.

When fixing locks on new doors and frames, a less unsightly but more secure keep plate can be used. Diagram 22 shows one example of a keep plate which has improved fixings. This is similar in design to the improved rim lock staple

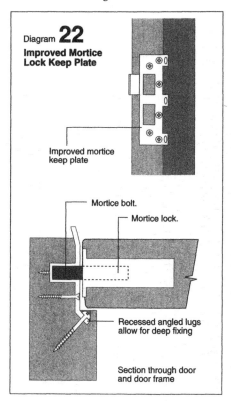

Diagram **22**

Improved Mortice Lock Keep Plate

Improved mortice keep plate

Mortice bolt.

Mortice lock.

Recessed angled lugs allow for deep fixing

Section through door and door frame

(see Diagram 16), having repositioned screw holes and angled lugs to provide a better distribution of fixings. There are other similar keep plates available.

Improving the door stile   The accuracy with which a mortice deadlock is fitted into the stile of a door can determine the performance of the lock against force. For example, if the width of the slot (mortice) cut to fit the lock is far greater than the width of the lock case, the combined resistance of the lock and door against force is unnecessarily reduced. In addition, hardwood door stiles tend to provide more resistance to force than softwood door stiles. Even where locks have been correctly fitted to doors, improvements will often be needed.

Fixing reinforcing strips or plates around the locking area of the door edge will more than compensate for the removal of timber to fit the lock. Diagram 23 shows two examples of mortice strips or plates. These are positioned on both sides of the lock and are bolted together through the stile of the door, clamping the timber and preventing splitting.

Maintenance of locking devices

Fortunately most locking devices usually require little in the way of general maintenance. When locks do become stiff or difficult to operate an application of

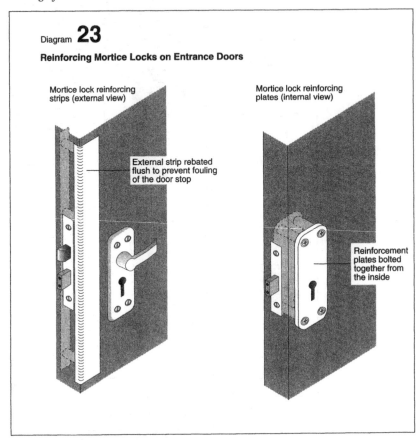

Diagram **23**

Reinforcing Mortice Locks on Entrance Doors

Mortice lock reinforcing strips (external view)

Mortice lock reinforcing plates (internal view)

External strip rebated flush to prevent fouling of the door stop

Reinforcement plates bolted together from the inside

light machine oil should be used for the workings of the lock, except for pin or disc tumblers which should only be lubricated with graphite. (Oil would clog up the mechanisms.) If the lock will not turn at all and the correct key has been inserted, then this would suggest a more fundamental problem such as misalignment of the door or a faulty lock mechanism.

5.6 LOCKS FOR FIRE ESCAPE PURPOSES

In many situations, the security of entrance doors needs to be balanced against fire safety requirements. This is particularly the case for entrance doors to flats in multi-storey blocks. This section provides guidance on securing doors to residential and non-residential buildings where fire regulations and the means of escape apply.

Entrance doors to flats in multi-storey blocks

The entrance door to a flat is usually the only means of entering or leaving the dwelling and ease of escape in the event of an emergency such as a fire is of critical

Diagram **25**

Mortice Escape Lock

Key operated
from outside

Inner snib or lever handle
to release and fasten lock

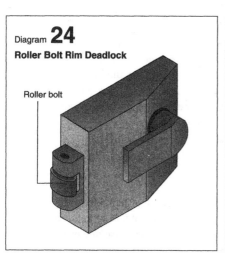

Diagram **24**

Roller Bolt Rim Deadlock

Roller bolt

importance. It is necessary that residents can escape quickly without having to hunt for keys to open exit doors. However, it is also important to provide a reasonable level of security so that residents do not feel the need to fit additional security locks and bolts. Such devices could easily hinder escape and their installation may undermine the fire resistance of the door. Locks considered suitable for entrance doors to flats in multi-storey blocks are detailed below.

Roller bolt rim locks

The simplest surface-mounted locking device to fix on to flat entrance doors is a rim lock which incorporates a roller bolt (an example is shown in Diagram 24). The roller bolt is designed to keep the door shut but allow it to be pushed open from outside or pulled open from inside. The roller bolt can be deadlocked (secured) with a key from outside the dwelling and deadlocked and released from inside by turning a thumb turn or handle. The main problem with these locks is the ease with which the lock and its staple can be ripped from the door and door frame. Although the staple can be strengthened by fitting reinforcing plates, or replaced with a proprietary brand staple designed with improved fixings, it is far more difficult to improve the fixing of the lock itself. Furthermore, the roller bolt may not have sufficient strength to keep the door shut in the event of fire. Where improvements involve passing steel bolts through the door or door frame, approval is needed from the building control officer.

Mortice locksets (escape)

These are a variation on the conventional mortice lock. The most common type
of lockset in use operates similarly to the rim lock described above, i.e. they are
operated from outside with a key and from the inside by turning a thumb turn or
handle (an example is shown in Diagram 25). An alternative lockset is the split
follower. Although this is deadlocked from both sides of the door with a key, the
door can still be opened easily from inside by turning the handle. In common with
all mortice locks, mortice locksets have weaknesses in the way they are fixed. Firstly,
the keep plate often needs to be strengthened (or even replaced) to prevent it from
being ripped from the frame. Secondly, the area surrounding the lock needs to
be reinforced to prevent the timber splitting and the lock dropping out. As with
improvements to rim locks, approval is required before installing hardware which
passes through the door or door frame.

Mortice locksets are available with either lever or cylinder mechanisms.

Multiple locking systems

Fitting two or more independent locks to fire doors is not acceptable to the
building control officers because it does not allow all the bolts to be released
simultaneously. There is also the risk that residents may only use one of the locks,
making it easier for an intruder to force entry. To overcome these problems, locking
devices have been developed which, at the turn of an inner handle, throw a number
of bolts simultaneously from the edge of the door into the door frame. Turning the
handle in the opposite direction will release the device. Being key activated from
outside the dwelling only, this kind of device should satisfy means of escape
requirements. Broadly, there are three types of multiple locking device used on
doors to dwellings:

1. The most suitable multiple locking device to fit to existing doors is the type
   fixed onto the inner surface of the door. An example is given in Diagram 26.
   The locking mechanism is usually fixed mid-way along the locking edge of the
   door. Rods or bars extend vertically from the mechanism in order to operate
   the upper and lower bolts. These devices are reasonably easy to fit and require
   little timber removal. However, in common with most surface fixed locks, their
   effectiveness is to a large extent dependent on the strength of fixing to the door.
   This is particularly the case for the keep plate (staples) fixed on the door frame.
   As a general rule, these devices should be fixed using the longest screw the
   timber can accommodate. This usually means penetrating the timber to about
   5 mm short of its width. Alternatively, nuts and bolts can be used to provide
   a stronger fixing than screws. Surface-fixed multiple locks may also be
   considered unsightly.
2. Multiple locks, as shown in Diagram 27 can be fixed into the locking edge of the
   door. It may possible to fit these to existing doors although this is a highly skilled
   operation and care needs to be taken not to reduce the integrity of the door or
   weaken its performance (against force or leverage). Furthermore, these locks are

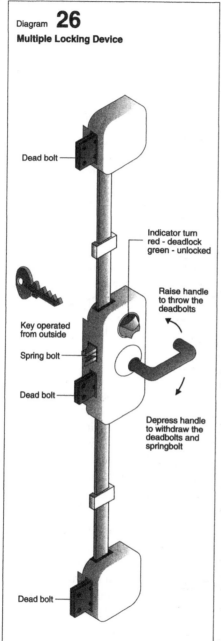

Diagram **26**
Multiple Locking Device

Dead bolt

Indicator turn
red - deadlock
green - unlocked

Raise handle
to throw the
deadbolts

Key operated
from outside

Spring bolt

Dead bolt

Depress handle
to withdraw the
deadbolts and
springbolt

Dead bolt

Diagram **27**
Multiple Locking Device

Top bolt engages
into head of door
frame

Key operated
from outside

Lever handle or
similar to fasten
or release all
the bolts

Bottom bolt engages
into sill of door frame

usually supplied with keep plates which extend the full length of the door frame, and prevents the fitting of intumescent strips. These locks should therefore only be fitted to doors which do not require intumescent strips in order to achieve their fire rating.

3. A variation of the device described above is the type which is fitted into the actual structure of the door. They operate in the same way and may even have bolts which are thrown into the head and sill of the door frame. Doors usually have to be specially made in order to accommodate this type of lock.

Note: Multiple locking devices require more maintenance than conventional locks. In particular, the bolts may require periodic adjustment to take account of wear. Failure to do so can often result in the misalignment of bolts and prevent residents from locking or even closing their doors.

Escape doors from non-residential buildings

Emergency exits from non–residential buildings such as community centres, youth clubs, etc., can be notoriously difficult to make secure. Often they are double doors located in secluded parts of a building, set back from the building line into recesses, and they open out into areas which are not overlooked or are seldom used. To facilitate ease of escape, they must be easy to open (without the need for a key) and must open outwards, which makes the hinges vulnerable to attack.

Panic bolts

Panic bolts are probably the most common emergency door release in use and are available for both single and double doors (Diagram 28). They consist of vertical bolts or rods which extend from the centre of the door and engage into keep plates at the top and bottom of the door. These are fixed to the inner surface of the door and operate by either applying a downward or forward pressure onto a horizontal bar spanning the width of the door. This action releases the bolts allowing the door to open.

Panic bolts have a number of inherent weaknesses:

- Rattling the door from the outside can loosen the bolts allowing the door to open. This can be a particular problem on panic bolts which are old or worn.
- They can be manipulated from outside by feeding a piece of string or wire under the door and attaching it to the inside horizontal bar. This can then be pulled to release the bolts. Furthermore, unless the meeting edge of double doors is rebated, wire can be inserted between the gap, and hooked onto the horizontal bar to pull the doors open.
- Dust, etc., can accumulate in the keep plate (socket) which is usually located in the floor. This will then prevent the bolt from engaging and may even cause damage to the bolts or the locking mechanism.
- Leverage applied to the door from outside could bend the bolts or even rip them from the door allowing the door to be pulled open.

Diagram **28**

Individual Bolts for Double Doors

FIRE EXIT   FIRE EXIT

Panic bolts need to be fixed securely to doors and be well maintained. The areas outside the doors should be clearly visible and well illuminated for both safety and security purposes. For vulnerable doors alternative devices should be considered. Consideration could also be given to installing a simple alarm system which would be activated if the doors are opened. This method is used extensively in shops, public buildings and offices. However, to be effective there is need for an immediate response.

Panic latches

Panic latches are similar to panic bolts except that instead of vertical bolts they have a latch which engages into a keep plate fixed to the frame on the side of the door.

Diagram 29
Barrier Bar.

FIRE EXIT

TO OPEN

Keep plate
preferably
bolted to
door frame.

This latch is housed in the mechanism which is attached to the surface of the door, or, on more expensive models, is morticed into the edge of the door. In either case, the security offered is dependent on how securely the device is fixed to the door and door frame. Panic latches are also susceptible to the problems described above for panic bolts.

Barrier bars

These are a heavy duty version of the panic latch and have bars which extend out both sides of the door into the frame (Diagram 29). They are designed primarily for single doors and are usually operated by turning a lever to withdraw the bolts. Barrier bars offer a higher resistance to leverage than panic latches.

Four point locking

This is a more robust version of the panic bolt door release system. It has four spring-loaded bars which extend out from the centre of the door into the head, sill

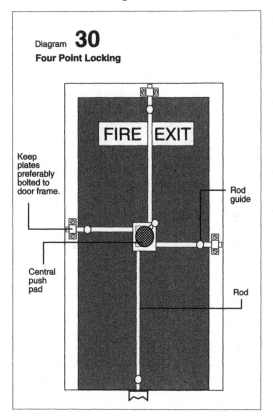

Diagram **30**
Four Point Locking

Keep
plates
preferably
bolted to
door frame.

Rod
guide

Central
push
pad

Rod

FIRE EXIT

and both sides of the door. It is released by striking a button in the centre of the door which causes the bolts to spring back and allow the door to open. This device, as shown in Diagram 30, provides a relatively high resistance for forced entry and is suitable for particularly vulnerable doors.

Key-operated locks

When a building is occupied, emergency exit doors should not require a key to open them from the inside. This includes the system whereby the key is contained inside a case with a glass cover which is designed to be broken in an emergency so that the key can be used to open the door.

Some emergency exit devices can be key operated from outside the building only if the door is also used as an entrance door.

Break glass systems

Certain door releases require glass to be broken in order for the door to be opened. These are designed primarily to discourage casual use of escape doors by

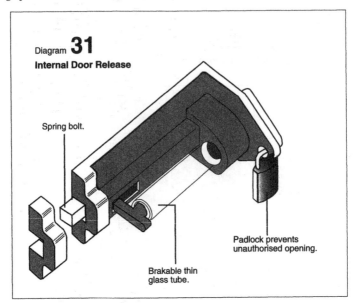

Diagram **31**

Internal Door Release

Spring bolt.

Padlock prevents
unauthorised opening.

Brakable thin
glass tube.

occupants of the building. One such device is shown in Diagram 31. This has
a thin glass tube which holds open a spring-loaded bolt. Once broken, the bolt
shoots back allowing the door to be opened. These devices offer only limited
security and should therefore only be used on internal doors such as those in
offices where the fire escape route passes through parts of the building occupied
by another tenant.

5.7　DOOR BOLTS

Entrance doors other than the door used as the final exit can be secured from
the inside of the dwelling. This means that less expensive locking devices, such
as bolts, can be used instead of additional locks. There are two basic types of
bolt: those fixed on the surface of the door and those concealed within the door
structure.

Improving surface bolts

Surface bolts, commonly known as slide, draw or barrel bolts, are fixed to the
surface of the door. All too often, bolts used on entrance doors are insubstantial,
providing little or no resistance to force. Bolts should be made of strong, robust
materials and fixed with long twin-threaded screws or, for greater strength, with
bolts and nuts fitted through the structure of the door.

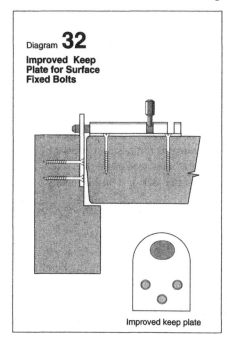

Diagram **32**

Improved Keep Plate for Surface Fixed Bolts

Improved keep plate

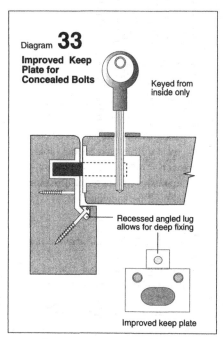

Diagram **33**

Improved Keep Plate for Concealed Bolts

Keyed from inside only

Recessed angled lug allows for deep fixing

Improved keep plate

Surface bolts can be further improved by replacing their keep plates with the type shown in Diagram 32. (These would need to be made as they are not commercially available.) This keep plate has improved fixings at right angles to the surface of the door, providing greater resistance to force.

Surface bolts are easily operated, encouraging their use. However, only key-operated surface bolts should be fitted if there is a possibility of access to the bolts from outside by mean of postal apertures or glazed panels.

Improving concealed bolts

Concealed bolts, also known as rack bolts, are fitted into the locking edge of the door. The bolt is shot into the surrounding frame by use of a common key from inside. The effectiveness of these bolts can be improved by fixing new keep plates as shown in Diagram 33. (These need to made as they are not commercially available.)

This is a similar design to the improved rim lock staple (see Diagram 16, page 57), with repositioned screws and angled lug.

Concealed bolts are both thrown and withdrawn by use of a key making them less vulnerable to manipulation from outside. On the other hand, there is a greater tendency for occupiers to neglect using them than would be the case with most surface bolts.

5.8 POSITIONING LOCKS AND BOLTS

All entrance doors to dwellings should have at least two lockable bolts which secure the door into its door frame. This can be achieved by fitting two separate locks to a door or by fitting multi-locking devices. Providing two separate locks on a door will inevitably be the least expensive option, particularly if up grading an existing door which already has a suitable lock which can be retained. It is unlikely that this approach will be acceptable to building control officers for entrance doors which are required as a means of escape such as those in multi-storey blocks. The reason for this is that, to open the door, all the bolts need to be released simultaneously. In these circumstances, a single escape lock or a multi-locking device should be specified.

Fitting locks and bolts to new doors

Where two locks are fitted to final exit doors, they should be positioned where it is anticipated that force may be exerted on the door. A suitable arrangement would be:

- A cylinder rim deadlock with improvements to the staple (see pages 56 and 57) should be fixed at around 1250 mm from floor level. This is roughly the height one would expect from a shoulder charge at the door.
- A mortice deadlock fixed at 650 mm from floor level. This is the anticipated height for a foot impact to the door. The fixing of the lock should be improved (see page 60) to increase its resistance to kicking and jemmying.

On less vulnerable doors which are locked from inside the dwelling, a mortice sashlock, improved as stated on page 57 and shown in Diagram 18, should be fixed at mid-height. In addition, two door bolts (see page 70) should be fixed at 150 mm from the foot and head of the door.

Fitting locks to existing doors

Where a lock is already fixed to a door, it is usually positioned at mid-height or above. The position of an additional lock should be mid-way between the floor and existing lock and, ideally, have a minimum distance of 400 mm between locks. This avoids the all too familiar practice of positioning the locks close together in one area of the door, thereby weakening the door stile further (by removing timber), making it easier to force the door.

Avoiding weakening the door

When installing locks, and in particular mortice locks, timber should not be removed where structural timbers of the door join the door stile. This would remove the majority of the joint and seriously undermine the strength of the door.

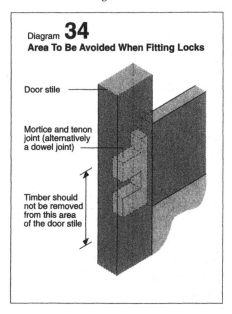

Diagram **34**
Area To Be Avoided When Fitting Locks

Door stile

Mortice and tenon joint (alternatively a dowel joint)

Timber should not be removed from this area of the door stile

Unfortunately, this is often common practice. Diagram 34 shows the area to be avoided in a typical 2XG door.

5.9 POSTAL APERTURES

Postal apertures in doors can provide opportunities for intruders to reach through and release door locks. This is particularly a problem for entrance doors to flats fitted with locks which open from inside without a key (for fire escape reasons).

When installing new postal apertures, it is important to consider carefully the position of the aperture and the size of the opening. Ideally, postal apertures should be located through a suitable external wall, away from the locking edge of the door. This may, however, prove impractical and expensive unless incorporated into the design of new buildings. Diagram 35 shows a simple design of a through-the-wall postal aperture.

An alternative to postal apertures is to provide external post boxes. This approach can have mixed fortunes. For instance, although they dispense with the need to have potentially vulnerable openings in or around doors, the post box itself could easily become a target. At one time, post boxes were a popular feature in small blocks of flats and were usually located in the internal entrance area close to the communal entrance door. However, many of these post boxes became prone to vandalism and theft of mail and, as a result, they have fallen into disuse. Careful consideration needs to be given to these issues if post boxes are to be used in housing developments.

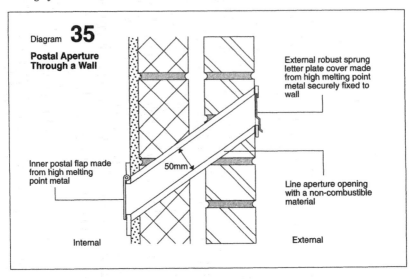

Diagram **35**

Postal Aperture Through a Wall

External robust sprung letter plate cover made from high melting point metal securely fixed to wall

Inner postal flap made from high melting point metal

50mm

Line aperture opening with a non-combustible material

Internal

External

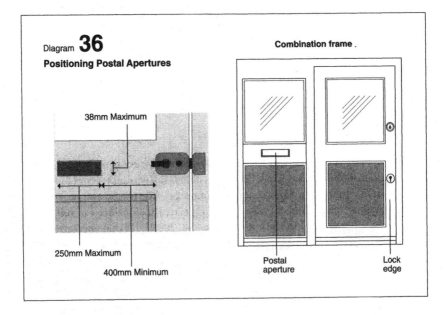

Diagram **36**

Positioning Postal Apertures

Combination frame .

38mm Maximum

250mm Maximum

400mm Minimum

Postal aperture

Lock edge

Postal apertures in doors

If the entrance door forms part of a combination frame, the postal aperture should be fitted in the frame and not the door. The aperture should be positioned as far away from the locking edge of the door as is practically possible.

If the postal aperture has to be located within the door, it should be positioned at least 400 mm from the door locks (see Diagram 36).

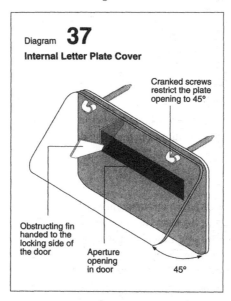

Diagram **37**

Internal Letter Plate Cover

Cranked screws restrict the plate opening to 45°

Obstructing fin handed to the locking side of the door

Aperture opening in door

45°

Wherever a new postal apertures is to be provided, the opening size of the aperture should be the minimum size recommended by the Post Office, which is 38 mm × 250 mm in accordance with the British Standard 2911: 1974 (1980): *Specification for letter plates*. Larger apertures increase the risk of intruders reaching through.

Improving existing postal apertures

Where existing apertures cannot be relocated, internal cover plates, letter boxes or baskets made from strong robust materials should be fitted to prevent locks being reached. These should be fixed securely to doors to ensure that they are not forced away in order to reach locks.

One of the key considerations when selecting cover plates, letter boxes or baskets is that they are designed to ensure that post, and in particular newspapers, can easily pass through the aperture into the dwelling and out of sight. There is no clearer indication that a dwelling is unoccupied than newspapers sticking out of postal apertures in doors. Letter boxes and baskets should be deep enough to take newspapers and prevent mail from being stolen. They can also be designed to allow mail to fall through and prevent accumulation during holiday periods.

Many entrance doors open into narrow entrance halls or directly onto internal walls which therefore do not provide sufficient space to fit internal letter boxes or baskets. Diagram 37 shows an example of one type of internal cover which would be suitable for these situations. This version of an internal cover plate incorporates a projecting fin which, when in the open position, prevents manipulation of the locks. This cover has also been designed with a restrictive opening of approximately 45 degrees. This has the advantage of reducing visibility into the dwelling through the aperture. These covers are handed so that the fin can be located on the side of the door where the locks are positioned.

5.10   DOOR VIEWERS AND DOOR CHAINS

This section deals with hardware which can be fitted to entrance doors to enable residents to identify callers without opening the door fully.

Door viewers

Door viewers should be fitted in circumstances where callers cannot be identified without opening the door; for example, to solid-core doors without vision panels or any other means of identifying callers. (It may be difficult to fit viewers in a suitable position on doors which have obscure glass panels.) The correct positioning of a viewer is essential to encourage its use. The viewer should be fixed at a height to suit the resident (which may seem obvious but it is frequently ignored). More than one viewer may be required to cater for the height of children, the elderly or disabled.

For clarity of vision, a viewer should have a glass lens and a wide angle view (160° minimum although 180° is preferred). A viewer should also have an internal lens cover to prevent callers from looking into the dwelling. It may be necessary to provide external lighting, such as small bulkhead lights next to entrance doors, to help identify callers. Lighting, where possible, should be controlled by a switch from within the dwelling or by a PIR sensor which activates the light if someone approaches the door. Any switches used to control lighting should be located within easy reach of the locking edge of the door to encourage use.

Conventional door viewers give a distorted 'fish-eye' image when looked through. Recent developments have produced door viewers which give an undistorted image. They work on the same principle as one-way mirrors, allowing the resident to see out but preventing a caller from looking in. They are larger than conventional viewers, and designed to be a feature of the door which can be engraved with the dwelling number. Although made from an extremely strong plastic (lexan), like all plastics they are difficult to clean without scratching the surface and could, over a period of time, reduce the clarity of vision.

Door chains

When choosing a door chain, emphasis is usually given to the strength of the chain. There are, however, two other important considerations:

- The strength of the fixings to both the door and door frame.
- The ease of engaging and disengaging the chain when the door is shut. The easier this is, the more likely it will be used.

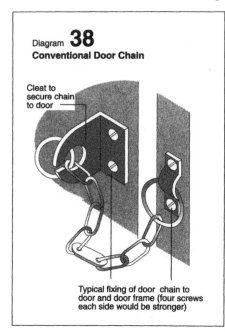

Diagram **38**
Conventional Door Chain

Cleat to secure chain to door

Typical fixing of door chain to door and door frame (four screws each side would be stronger)

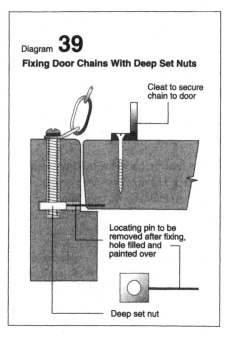

Diagram **39**

Fixing Door Chains With Deep Set Nuts

Cleat to secure chain to door

Locating pin to be removed after fixing, hole filled and painted over

Deep set nut

The usual method of fixing chains to door frames is to use wood screws. Diagram 38 shows a typical chain fixed with two screws into the frame. Door chains with four fixings into the door frame are preferable. The screws should be at least 50 mm in length. Door chains of this type should be fixed directly on to the doorframe, and not through cover mouldings such as architraves.

A more effective method of fixing a chain to the door frame is to use a deep set nut and bolt, as shown in Diagram 39. This chain also has far stronger screw fixings into the door.

5.11 INTERNAL DOORS

Locking internal residential doors, particularly with key-operated locks, can be counterproductive and may even be dangerous for the following reasons:

- When a dwelling is occupied, locked internal doors, especially those on the first floor or above, can hinder escape routes in the event of an emergency such as a fire.
- Locked internal doors, in particular hollow flush type doors, can easily be damaged by an intruder.

However, in some circumstances residents may feel safer if certain internal doors (such as doors in living rooms or bedrooms) are fitted with locking devices. If this is the case, simple devices such as door bolts or low security locks can be used. These devices should only be fitted and operated from within the room. This will prevent them from being used when the dwelling is empty.

Key-operated devices should not be used. It is best to use devices which slide to fasten and release, or locks which operate using a thumb turn, similar to those used on bathroom doors.

5.12   CHECKLIST FOR ENTRANCE DOOR SECURITY

The following checklist has been compiled for easy reference to enable planners, designers and maintenance personnel to identify weaknesses and decide where improvements should be made. The checklist should be used when installing new doors and frames as well as for upgrading existing ones. For ease of use it is cross-referenced with the relevant pages of this book.

1. Where site details allow, locate entrance doors where they can be easily observed. If this is not possible, assess the level of protection required. (See Location of door, page 45.)
2. The door should be strong and made from robust materials. (See Solid-core/ doors', page 48.)
3. Glazing within doors or surrounding frames must not be easy to break, giving access to locking devices. (See Fully glazed doors, page 47; Partially glazed doors, page 48.)
4. Avoid using flimsy panelling in both the door and surrounding door frame. (See Timber panels in doors, page 48; Partially glazed doors, page 48.)
5. Ensure that door frames are fixed securely to the surrounding superstructure. (See page 49.)
6. All entrance doors should be hung on at least three hinges and reinforced with two hinge bolts (three hinge bolts on outward-opening doors). (See Improving conventional hinges, page 51.)
7. Fit two key-operated locks to final exit doors, fixed in the appropriate positions. Ensure that they are fixed securely to both the door and door frame. (See Fitting locks and bolts to new doors, page 72; Fitting locks to existing doors, Avoiding weakening the door, page 72; Improving the performance of cylinder rim locks, page 56; Improving the Performance of mortice locks, page 59.)
8. All other entrance doors should have at least one key-operated lock and two good-quality bolts, preferably key operated. (See items mentioned in 7 above, and Door bolts, page 70.)
9. If possible, locate postal apertures well away from doors, particularly their locking edges. Make improvements to existing vulnerable apertures. (See Postal apertures, page 73.)
10. Where callers cannot be identified without opening the door, door viewers should be fitted. (See Door viewers, page 76.)
11. Fit door chains or limiter bars to entrance doors at which visitors are likely to call. (See Door chains, page 76).

12. Where necessary, provide external lighting to entrance doors to help identify callers. (See Door viewers, page 76.)
13. Ensure that residents fully understand, if necessary by demonstration, how and when to use security devices.

5.13 THE TERMINOLOGY OF LOCKS

This section has been compiled to explain many of the technical terms and functions associated with locking devices.

Anti-drill plates Toughened steel plates which surround the lock cases, usually fitted to less expensive mortice locks. Their purpose is to prevent the lock mechanisms being damaged or destroyed by cutting tools such as electric drills. However, the risk of this happening is very low. It is easier to drill out the timber surrounding the lock.

Back plate A metal plate bolted and/or screwed to the surface of the door on which to mount a rim lock.

Back set This is the horizontal distance between the outer face of the lock (usually the fore-end) and the centre of the key hole. This distance will vary depending on the lock but there is a standard distance applying to most locks. The back set will determine the position of the key hole on the door.

Dead bolt This is the square-ended bolt of a lock which is moved in and out of the lock case by a key. Once locked it cannot be pushed back into the case (unlocked) without the key. Some dead bolts are available which are operated from the inside by a thumb turn or knob only. These are designed specifically for ease of escape, e.g. on doors to flats in multi-storey blocks.

Dead latch Similar in shape and function to a spring latch (*see* Spring latch) but has the facility to be deadlocked (*see* Deadlocking). Those fitted with an auxiliary bolt will deadlock automatically when the door is shut.

Deadlocking This prevents the bolt from being pushed back into the lock case without the use of a key, thumb turn or knob. It prevents manipulation by a credit card or steel tape. All new locks fitted to entrance doors should deadlock.

Differs An abbreviation of 'differed combinations'. The larger the number of different combinations of keys available for a particular lock, the less the risk of someone having the same key to operate it. Specifying the number of differs is more important than specifying the number of levers on a mortice lock.

Double forehand A secondary plate, fixed to the forehand for decorative purposes (see forehand); available in a range of finishes.

Double throw Some locks have the facility to extend the dead bolt further into the keep/striking plate. This is usually by inserting the key in the mechanism and rotating two full turns in the opposite direction to that of releasing the lock. Theses locks are particularly useful on double doors which do not have rebated meeting edges.

Escutcheon A cover for the key hole of a mortice deadlock fixed on the surface of both sides of the door. Some external escutcheons have decorative secondary covers over the key hole. External escutcheons are also available for use with cylinder rim locks.

Forehand The exposed face of a lock, usually a mortice lock, through which the bolt(s) protrude, and by which the lock is fixed to the door.

Hardened steel rollers These are toughened steel bars (usually two in number) which are inserted horizontally into the bolt during manufacture. Their purpose is to increase the time it would take to cut the bolt using a hacksaw or similar cutting tool.

Latch A lock such as a sashlock which has a spring bolt (*see* Spring bolt); operated from both sides of the door using lever handles.

Lever Levers are a type of lock mechanism. They are contained within the lock case and have to be lifted a precise distance by a key in order for the lock to work. Mechanisms with a number of levers, each having to be lifted a different distance, provide protection against manipulation by an incorrect key. Locks used on final exit doors should have a minimum of five levers and, more importantly, a minimum of 1000 differs (*see* Differs).

Lever handles These are horizontal handles fixed on the surface of both sides of the door. Applying downward pressure opens the spring latches. On release they return automatically to the horizontal position.

Lock case This contains the lock mechanism. Mortice lock cases are usually made of hardened steel to prevent the mechanism from being damaged or destroyed by cutting tools such as electric drills.

Locking edge The edge of the door on which the locks are either mounted or concealed.

Mechanism The working parts of the lock and the way those parts perform to achieve the required level of security. Main types include pin and disc tumbler or lever. More specialised types are available for greater key security for use on safes, etc.

Mortice A slot cut into the locking edge of a door in order to receive a mortice lock.

Mortice lock A general name used to describe a variety of locking devices which are concealed within the locking edge of the door. Usually the function of the lock is included in its description, e.g. a mortice lock which has a dead bolt is known as a mortice deadlock.

Mushroom driver   An effective anti-picking device incorporated within cylinder mechanisms. The driver is tapered and has a mushroom-shaped head which prevents manipulation of the lock with a lock pick or similar implement.

Night latch   A rim or mortice lock which has a spring bolt. It is usually operated from outside by a key and inside by turning a thumb turn or lever handle. It should not be used in situations where it is easy to gain access from outside through glazing, postal apertures, etc.

Pin tumbler mechanism   A lock mechanism incorporated within a cylinder fixed through the structure of the door. The key is cut with a series of 'V' notches on the top edge which, after exerting the mechanism, correspond to a series of pins and drivers. The correct key lines up the pins allowing the key to operate the lock. Five pin tumbler mechanisms offer up to 24,000 differs as standard, providing a high level of protection against false keys. Disc tumbler mechanisms are similar except they have disc instead of pin tumblers.

Rim lock   A type of lock fixed on the inside surface of the door. Various types are available, some having spring bolts which can be deadlocked.

Roller bolt   A spring bolt which is made in the form of a roller. This keeps the door shut without needing to use a key, lever handles or thumb turn but still allows the door to be pushed or pulled open. The roller bolt can be deadlocked from outside by using a key and from inside by a thumb turn. They are used extensively on entrance doors to flats in multi-storey blocks.

Sashlock   A mortice lock which incorporates both a dead bolt and spring latch. Usually the dead bolt is key operated both sides of the door although some are available with thumb turns on the inside only. The latter are used on doors to flats in multi-storey blocks.

Spring/latch bolts   A bolt with the outer edge bevelled on its vertical face. It can be pushed back into the lock case and will return to the extended position without mechanical assistance. It automatically locks when the door is shut. It is opened from the outside by a key and from inside by a thumb turn or knob.

Spring latch   Similar to a spring bolt but opened both sides of the door by lever handles.

Staple   Fixed on the inside of the door frame, a box-like fitting which receives the spring bolt of a rim lock when the door is closed. If used on inward-opening doors, it is a vulnerable part of the lock which often needs up-grading.

Striking/keep plate   A shaped metal plate fixed on the inner face of the door frame. It usually has one or two holes, depending on the lock, into which the dead bolt and/or spring latch engages. It is used for all mortice locks and rim locks which have had their spring bolts reversed for use on outward-opening doors.

Thumb turn or knob   Used internally, a thumb turn or knob is a small handle fitted to the lock in order to operate the spring bolt or dead bolt without the need of a key.

WINDOW SECURITY

5.14   VULNERABILITY OF WINDOWS

It is estimated that over half of all household burglaries involve entry through windows, and mainly those situated at the rear of the dwellings. In fact any window which is readily accessible is potentially vulnerable, more so if it is secluded and out of sight from passing pedestrians, motorists and neighbours. To achieve a reasonably good level of security for dwellings, all accessible windows need some form of security protection.

The level of protection required will vary, depending on the location of the window and the design of the frame.

Location of window

As a general rule, windows situated at the rear or to the side (especially those in communal passageways) of dwellings, on the ground floor or in basements, or those in secluded positions close to flat roofs, drain pipes or fire escapes, are most at risk. The nearness of adjoining balconies and boundary walls can also give access to first-floor windows or above.

Window frame design

The design of the window frame and its glazing can be a factor in determining its vulnerability. For example, a well-maintained sash frame can be notoriously easy to force open unless it is fitted with an appropriate lock. The physical condition of the window (including glazing) and its fixing to the surrounding structure are also important considerations.

Furthermore, windows with larger panes of glass tend to be more secure. This is because they are slightly more difficult to break than smaller panes of glass and the amount of noise caused when they are broken is likely to attract unwanted attention.

5.15   FIXING WINDOW FRAMES

Having ensured that the window frame is in good enough condition to be retained, it is important that it is fixed securely to its surrounding structure. It is usually relatively easy to refix a timber window frame with fixings specifically designed for this purpose. These are generally known as frame fixings and consist of a long plastic plug and steel screw. These are inserted into a single hole drilled through the frame and into the surrounding structure.

Aluminium and steel frames are usually installed into outer timber frames and can therefore be treated in the same way. Alternatively, steel frames can be held in place with lugs which were welded to the frame and built into the surrounding structure during the construction process. Over time, these lugs may corrode away leaving the frame loose. The fixing method described above for timber frames can usually be used for steel frames unless the corrosion has eaten into the frame itself. In these circumstances the frame will probably need to be replaced.

uPVC frames are generally installed using a bedding compound round the edge of the frame and, once set, holds the frame in place. It also helps to seal any gaps or holes between the frame and the surrounding structure. It is also good practice to supplement the compound with frame fixings as described above.

5.16 VENTILATION

Ventilation into dwellings is essential where solid fuel is used for heating and to prevent problems of condensation. Small opening fanlights were incorporated into window frames for this purpose. However, these fanlights present a particular security risk, especially where they are located above much larger opening casement frames. The glazing in them can be easily broken (without creating too much noise) in order to reach handles of the larger opening frames. They are also frequently left open when dwellings are left unoccupied.

Another problematic method of providing ventilation is to lock the window in an open position, usually with a gap between the frames of around 10 to 15 mm. In order to do this a specific type of lock is required. This approach is common on uPVC frames. The main problem with this is that it is far easier to force open a window that is already partly open (even if it is locked) than one that is closed.

A far more secure method of providing ventilation is to install trickle ventilators, of which there are two main types:

- those that are permanently open – for use where constant ventilation is required such as in kitchens; and
- those that are adjustable i.e. can be opened or closed depending upon ventilation requirements.

Most trickle ventilators (Diagram 40) are fixed into the horizontal rail of the window frame which is deep enough to accommodate them, as shown in Diagram 41. Existing frames have, in many cases, rails that are too narrow for this purpose.

Simple plastic fan ventilators or electric extractor fans can be installed in holes cut in the window glazing to provide ventilation. However, these can be pushed or pulled through the glass to give access to the window catches. If extractor fans are needed, they should be installed through an external wall and well away from the windows.

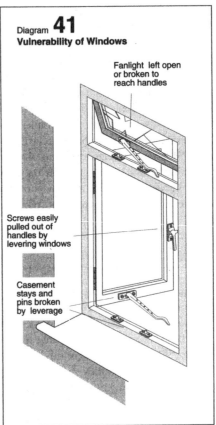

Diagram **41**
Vulnerability of Windows

Fanlight left open
or broken to
reach handles

Screws easily
pulled out of
handles by
levering windows

Casement
stays and
pins broken
by leverage

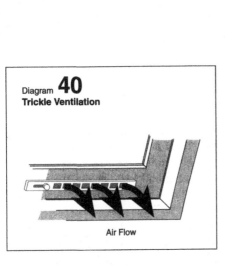

Diagram **40**
Trickle Ventilation

Air Flow

5.17   WINDOW LOCKING DEVICES

There is a common view that fitting locks to windows will do little to prevent
burglaries because intruders only need to break the glass and climb through.
Evidence suggests, however, that most intruders will not climb through broken
glass in windows. If glass is broken it is usually in order to reach in and release the
window catches. Alternatively, access is gained by forcing open the frame through
leverage of vulnerable window catches. Examples of these problems are shown in
Diagram 41.

Clearly, the risk of entry through all accessible windows can be reduced by
installing suitable window locks.

Types of window lock

A visit to a well-stocked architectural ironmongers will reveal that a considerable
range of window-locking devices is available. This is due partly to the wide range

of window designs and the different materials from which they are made as well as the costs of each type of lock. It is also likely that, for common window designs, there will be a number of locks available. When these considerations are combined, it can make the selection of the most suitable lock for a window difficult. In an attempt to address this, the following are descriptions of the main types of window lock available, highlighting their advantages and disadvantages.

Locks fitted to existing window ironmongery

There are ranges of locks that have been designed to be fitted on to existing window ironmongery such as handles and stay bars, as shown in Diagram 42. These tend to be the least expensive window locks but their effectiveness relies on:

- the strength of the existing ironmongery; handles and stay bars made of brittle materials such as aluminium or cast iron can easily snap if leverage is applied to the opening frame:
- the fixing of existing ironmongery; often handles and stay bars are fixed with small screws which can easily be pulled out if the window is forced.

For certain types of window frame, there may be no alternative but to fit locks on to existing ironmongery. For example, this is the case for some metal windows where the narrowness of the frame prevents the installation of other types of lock. If, in these circumstances, the frames have any of the problems described above, then other methods of improving window security should be considered.

Independent locks

As the title suggests, these are locks which are fitted independently from existing window ironmongery, as shown in Diagram 43. Inevitably these are more expensive but afford better protection as they are more robust and can be fixed more securely to the window frame.

Lockable window ironmongery

These are handles and stay bars which have in-built locking mechanisms and are typically fitted to new window frames. Once again the fixings of the ironmongery are critical to the performance of the lock.

Locking methods

The locks described above are all operated by one of three locking methods.

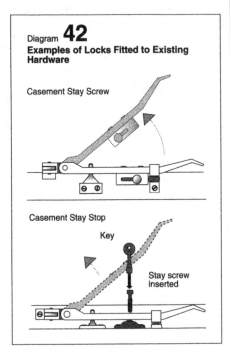

Diagram **42**
Examples of Locks Fitted to Existing Hardware

Casement Stay Screw

Casement Stay Stop

Key

Stay screw inserted

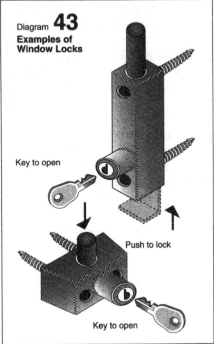

Diagram **43**
Examples of Window Locks

Key to open

Push to lock

Key to open

Manual locking

This method involves the use of a key to lock or unlock the window. These usually form the least expensive range of locks and rely on a locking screw thread to fasten the window. The main advantage with this type of device is that as the screw is tightened (i.e. to lock the window), it pulls the opening frame tight against the window frame making it difficult to force open with a jemmy (or similar implement) from the outside. This is particularly useful if the opening frame is bowed or warped. It is important to recognise, however, that manual locks can quickly become a nuisance, especially if there are a large number of windows to lock. A typical house can have as many as 12 or more accessible windows that need to be locked. Manual locks could easily fall into misuse, especially if fitted to windows that are in constant use such as kitchens and bathrooms.

Semi-automatic locking

In this method locks are fastened by pressing a leaver or bolt but require a key for unfastening. These are more expensive but have the advantage of being far easier to operate and therefore less likely to fall into disuse. Furthermore, because the bolt or leaver protrudes from the lock when it is open, it is easier to see if the lock is fastened or not, as shown in Diagram 43.

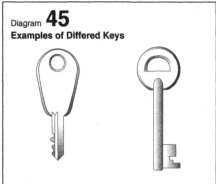

Fully automatic locks

These fasten automatically when the window is closed and require a key to open. In terms of encouraging residents to use window locks, these may appear to be the ideal solution. There are, however, drawbacks. Problems could arise if the opening frame warps or bows away from the window frame thereby creating a gap. This could easily prevent the lock from fastening when the window is closed. Positioning locks close to handle or stay bars should help to reduce this problem. Another difficulty could emerge if these are fitted to large frames where two locks may be required. In these circumstances both locks would need to be opened with a key simultaneously, as well as the window itself.

Window lock keys

Window locks use two types of key:

1. Common keys, which fit every lock in a particular range. This means that it is possible to operate all the window locks in a dwelling with the same key. There are different types of common key, see Diagram 44, which are usually unique to a particular manufacturer. While this is a convenient system, it does mean that the keys are more widely available. Their relatively simple design also means that they can be opened with other implements, such as a small screwdriver for example. Window locks using common keys are really only suitable for securing one-off dwellings where the chance of dwellings in the area having the same type of window lock is small. They should not be specified for use across the whole of a housing estate.
2. Differed keys as shown in Diagram 45 offer a variation of keys in a lock range. These use a smaller version of the cylinder keys used on rim locks for entrance doors, although they do not have as many key variations. It is still possible to have all the locks in a dwelling operate on the same key by matching or suiting the locks. Additional keys for these locks have to be cut from the original which obviously benefits building security. They are also more suitable for use where window locks are to be fitted to a number of dwellings in an area.

At one time a window lock was available which did not need a key to operate at all. Instead it had a combination lock (similar to those used on briefcases) of three numbers which had to be aligned in a predetermined sequence before the lock could be opened. The combination could be changed as and when necessary. This lock had the obvious advantage that there are no keys to be lost or stolen. It also overcame one of the biggest problems with window locks, that of keys being left in the locks. Unfortunately these locks were rather expensive and have not been widely used.

5.18   WINDOW FRAME DESIGNS

This section is concerned with selecting the most appropriate type of lock for particular window frame designs and identifying the most suitable position at which to fix the lock with the aim of giving the best possible security performance. Each type of window will be dealt with separately. However, as a general rule, the position of locks will depend on:

- the direction in which the window frame opens;
- the size of the opening window frame; two locks may be required on large opening frames of 900 mm along any one side of the frame;
- obstructions – such as sink units in front of kitchen windows – to internal access to the frames.

Casement frames

Diagram 46 shows the position of locks for a typical side-hung casement frame with opening fanlight. It is important to fit locks to all opening frames if they are readily accessible from outside.

Sash window frames

Sash windows can be notoriously easy to open from outside by sliding the blade of a knife (or similar implement) between the two horizontal meeting rails and forcing back the window catch. This problem can be overcome by replacing the catch with a lockable cam (or fitch) catch, as shown in Diagram 47. Alternatively, a pair of dual screws can can be fitted to lock the two opening sashes together when in the closed position. This is shown in Diagram 48.

Sash windows are often used in dwellings and other buildings which open directly onto public footpaths because, when open, they create no hazard to passing pedestrians. However, this can create a security risk if windows are left open in unattended rooms, particularly on warm summer days, or in rooms which require constant ventilation, e.g. kitchens. This risk can be reduced by:

- sealing the lower sash into its frame, to prevent it opening, and using the upper sash for ventilation;

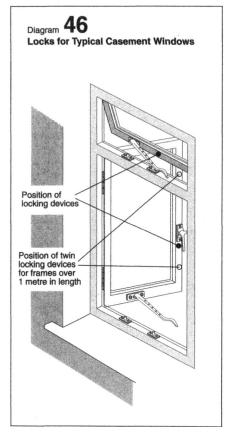

Diagram **46**
Locks for Typical Casement Windows

Position of
locking devices

Position of twin
locking devices
for frames over
1 metre in length

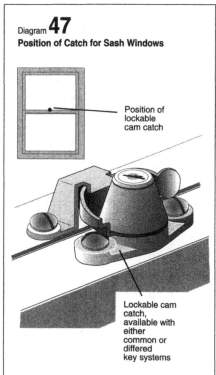

Diagram **47**
Position of Catch for Sash Windows

Position of
lockable
cam catch

Lockable cam
catch,
available with
either
common or
differed
key systems

■ fitting lockable window stops which limit the distance both sashes can open; as shown in Diagram 49, the opening distance should be no more than 150 mm.

Note: Window stops can be weakened by a determined intruder by sliding the window sash up and down in a battering ram fashion. Therefore they should not be left open when the dwelling is unoccupied.

Horizontal sliding windows and doors

Horizontal sliding windows are similar in design to sliding patio doors and therefore have the following common problems.

1. The material used to form the opening frame may be too flimsy to accept a locking device. In these circumstances, either the existing frame should be replaced with a more robust frame, or internal grilles should be considered.
2. Horizontal sliding windows and patio doors can also be a security risk if the sliding glazed panels can be lifted out of the outer window frame. To help

Diagram **48**
**Position of Dual Screws
for Timber Sash Windows**

Position of
dual screws

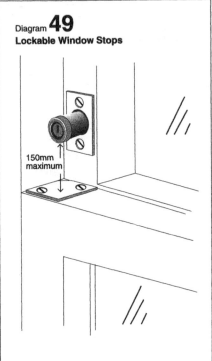

Diagram **49**
Lockable Window Stops

150mm
maximum

prevent this, a packing timber strip can be placed between the head of the sliding frame and the outer frame, as shown in Diagram 50.

It is important to note that patio doors tend to be targets for handy missiles (such as plant pots) left lying around in back gardens. Laminated glass should be used as both a safety and security measure.

Louvre windows

The guidelines offered so far have been related to window panes fitted within a casement or sash. Their security can generally be improved by installing window locks. Louvre windows, however, present an additional security risk because there are a large number of opening glass panes which, in most cases, are notoriously easy to slide even when the window is closed. Louvre windows also tend to cause draughts.

The security performance of louvre windows can be improved to some extent by:

- defacing or replacing exposed fixing screws with non-returnable screws;
- replacing the existing glass with laminated glass which is then fixed in place with adhesive. (This does mean that broken glass is more difficult to replace if damage occurs.)

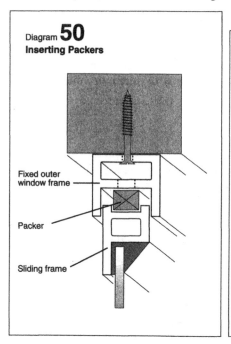

Diagram **50**
Inserting Packers

Fixed outer
window frame

Packer

Sliding frame

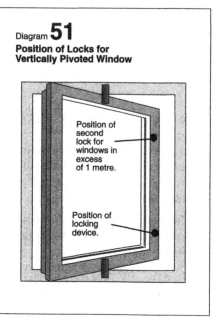

Diagram **51**
**Position of Locks for
Vertically Pivoted Window**

Position of
second
lock for
windows in
excess
of 1 metre.

Position of
locking
device.

To provide a better level of security, louvre windows should be either taken out and replaced with a different type of frame, or protected by internal window grilles.

Pivot frames

Pivoted windows can open either vertically or horizontally. As a general rule, window locks should be fitted on the side of the frame which opens outwards. This is shown for vertically pivoted windows in Diagram 51.

Pivoted windows can have their hinges located on the outside of the frame, which means that they can be vulnerable to tampering. Although this is not regarded as a common problem, the hinge-fixing screws could be either defaced or replaced with non-returnable screws.

Skylights

Skylights can be provided in both pitched and flat roofs. It is unlikely that skylights in pitched roofs will present a security problem because of limited means of access. Skylights located in flat roofs can, however, cause serious security problems. Even skylights which are constructed from robust materials such as polycarbonate and are strongly fixed may still be a security risk.

The only effective way of providing an adequate level of security for accessible sky/roof lights is to fit internal grilles within the well of the opening. The grille should be made of an adequate gauge of steel to resist bending and be securely fixed using bolts or similar robust fixings. The length of the bolt will depend on the thickness of materials used in the construction of the roof opening. As for most fixings, use the longest bolt possible.

Note: Care needs to be taken when fixing grilles to skylights to ensure that waterproof membranes are not pierced, allowing water to enter into either the building or the structure of the roof.

5.19   WINDOW LOCK INSTALLATION

Experiences from security improvement schemes and projects have shown that careful planning of window lock installation is required, especially if a large number of dwellings are involved. Forward planning should take the following considerations into account.

1. The condition of the window frames. This should include the frames themselves, the hinges, glazing material, window hardware and glazing compounds (e.g. putty).
2. The window frame should be securely fixed to the surrounding superstructure. For improved or additional fixings, see section 5.15.
3. Ideally one type of lock should be used in each dwelling. The type chosen should be compatible with the design of the window. Where there are two or more designs of window in a dwelling, choose from a single manufacturer who can provide interchangeable keys, so that the same key will operate all locks in the dwelling.
4. Where locks are installed to a number of dwellings in one area – for example, a whole estate – it is advisable to use locks which have a 'differed' key system. (See Window lock keys, page 87.)
5. Particular attention should be given to vulnerable windows. Extra precautions may be required over and above the installation of window locks (see section 5.14).
6. The choice of window locks should be flexible to take account of special needs, e.g. a person with arthritic hands may have difficulty using locks which have small keys. This is a real problem as most window locks with a differed key system are supplied with small keys.
7. Clear instructions need to be given to window lock installers regarding the correct positioning of the lock on windows (for each size and type of window). Instructions for residents with special needs must also be given.
8. Clear operating instructions should be given to residents on using window locks. It may also be necessary to show residents how to use them. Attention should be drawn particularly to the importance of removing keys and keeping them in a safe, but known, place and not to take them out of the dwelling. This will only increase the risk of losing keys or having them stolen.
9. Effective maintenance is essential to ensure the long-term effective use of window locks. One of the main problems which may arise, particularly on timber windows, is movement within the frame due to climatic conditions. After the initial installation of the locks it may be necessary to return and make adjustments to take this into account.

5.20 GUIDE TO GLAZING MATERIALS

There is a variety of glazing materials available, each with differing performance characteristics. Listed below are the main types of glazing materials in general use, along with a description of their characteristics and applications.

Annealed glass

Annealed glass is by far the most common glass in use and is available in the following forms: float, sheet, cast (patterned), wired. The strength of annealed glass depends upon its thickness, overall size and, to a certain extent, its age (it becomes more brittle as it gets older).

The advantages of annealed glass are that it is readily available and can quickly and easily be replaced at relatively low cost. With the exception of wired glass which creates a dull sound when broken, annealed glass creates considerable noise when it is broken, especially in larger panes. However, wired glass will act as a barrier even when it is broken and this will continue until the wire itself is sheared.

Annealed glass is relatively easy to break and, once broken, produces sharp splinter-like shards which can easily cause injury. It should not therefore be used in low-level glazing (below waist height) or glazing within doors or adjacent door frames. For vulnerable windows and doors, alternative glazing should be used.

Toughened or tempered glass

This is float or sheet glass which has been subjected to a heating and cooling process designed to increase its surface strength. It can be between three and five times as strong as ordinary annealed glass. Although toughened glass is resistant to fracture by blunt instruments, it can be shattered by sharp objects. It also needs to be made to measure and therefore takes longer to replace. Once broken, toughened glass will no longer act as a barrier.

Toughened glass should be used as safety glass rather than for security (e.g. glazing within internal doors, etc.).

Laminated glass

Laminated glass consists of a sandwich of two or more sheets of ordinary glass bonded together with interlayers of plastic, polyvinyl butyral (PVB). Depending on requirements, for residential applications two categories are available – 3 ply, consisting of two sheets of glass with one interlayer of plastic, or 5 ply, having three sheets of glass and two of plastic.

The advantage of laminated glass is that, although the actual glass can be broken, the plastic interlayers will hold the glass in place; it therefore continues to act as a barrier making further penetration of the window more difficult. Compared to ordinary glass, laminated glass is more expensive. It is, however, particularly useful for protecting vulnerable windows and is used for both safety and security glazing in entrance doors and frames surrounding doors.

Plastic glazing

Plastics are increasingly being used for glazing, particularly in vulnerable areas such as entrance foyers in multi-storey blocks. The performance of plastics varies according to thickness, polymer type and method of fixing to the window frame.

Polycarbonate is often regarded as the the most suitable plastic for security glazing because of its high resistance to breakage. However, in common with most plastics, its surface can be easily scratched or defaced, even by normal cleaning operations, and improvements such as plastics with scratch-resistant surfaces have been made to reduce this problem. Furthermore, it will easily burn which may cause problems with fire regulations.

Obscure (patterned) polycarbonate could be suitable where light is important but vision is not, such as in youth clubs, community centres, etc. Clear polycarbonate is also suitable for use on vulnerable windows as a secondary glazing material where the existing glass and frame is to be retained. It should be fixed on the inside of the window so that the existing glass will protect it from the elements and minor abuse. Also, should the glass be broken, the polycarbonate will make further penetration of the window more difficult.

Double or secondary glazing

The primary purpose of double or secondary glazing is to reduce heat loss and noise penetration, but it also has a security value. The additional layer(s) of glass increase the difficulties of penetration as long as the casement/sash is secured into the frame and the frame is securely fixed. Enhanced security is achieved by using plastic for the second layer of glazing. Plastic should be fitted to the inside of the frame, with the glass to the outside; this protects the plastic from the elements and external defacement.

Polyester film

This is a thin layer of plastic film which is applied to the inner surface of existing glazing to increase the performance of the glass. The purpose is to produce similar characteristics to that of laminated glass, i.e. if broken, to hold the glass in position. However, it is not as effective as laminated glass. Its impact resistance is low, its surface can be scratched easily and, as it is placed on to existing glass, it is not fixed to the window frame thus reducing its overall effectiveness. Furthermore, it will not withstand a series of blows from a more determined intruder. The film will therefore need to be replaced from time to time.

5.21   FIXING GLAZING MATERIALS

Removing putty or glazing beads in order to take out glass as a means of gaining entry is not common. However, this risk can be further reduced.

1. Replace putty, which cracks and shrinks with age, if it is in poor condition. Replacing with new putty should be limited to windows which are not particularly vulnerable as putty is easily removed before it sets hard. Wherever possible, particularly vulnerable windows should be fitted with glazing beads, but this may only be practical for timber-framed windows.
2. Glazing beads need to be properly secured in place. When improving existing glazing beads or fitting new ones, the following guidelines should be followed:
   - Screws should be used to fix beads to frames and should not be more than 200 mm apart. Screws should be no less than SG 8.
   - The length of screw will depend upon the thickness of the glazing bead. However, penetration into the frame should be a minimum of 32 mm.
   - Screws should be countersunk, stopped up with filler and painted over to disguise the points of fixing.

Plastics

Using plastic as glazing material presents a different problem. Due to their flexibility, plastics in larger windows can be sprung from their frames unless they are fixed securely.

For timber windows, round-headed wood screws can be used to fix the plastic into the glazing rebate of the frame. Fixing into metal windows can be more difficult, depending upon the material used for the frame. Aluminium frames are usually flimsy and the only reasonably secure way of fixing plastics to them is by using rivets. Steel windows tend to be more robust, and self-tapping screws can therefore be used.

Whichever fixing method is used, ensure that there are no exposed screw heads. Sealants are also required to seal plastic into the frame to prevent moisture or water penetration.

5.22 PROTECTING VULNERABLE WINDOWS

Particularly vulnerable windows, such as those in basement dwellings, community buildings and neighbourhood offices, may need additional security over and above the measures detailed so far. Much will depend on the type and location of the window. These extra improvements could include some form of grille.

Window grilles

Window grilles can vastly improve the security of a window. Unlike other barriers such as laminated glass, they are obvious visible barriers which will often deter an attempt to force entry.

Grilles in the form of 'prison bars' or wire mesh greatly detract from the appearance of a building and are particularly out of place in dwellings. Fortunately, it is possible to produce window grilles which are reasonably attractive yet provide a

Plate 11 Example of a decorative window grille

good level of security (as in Plate 11). Window grilles should be fixed on the inside of a building rather than the outside. This will considerably reduce corrosion, but, more importantly, it will prevent tampering with the fixings.

Wrought iron type grilles

Wrought iron type grilles can be produced in virtually unlimited designs such as scroll patterns, etc. However, for security reasons, they should be designed with the following elements incorporated:

1. The outer frame of the grille should be made with an adequate gauge of steel to ensure rigidity. The gauge required will depend on the size of the opening but a minimum of 20×6 mm is required.
2. The gauge of steel which forms the design within the frame should be a minimum of 13×3 mm.
3. Spaces created within the grille should not exceed 200 mm^2.
4. Grilles for larger windows (above 1,500 mm^2) should have intermediate horizontal and/or vertical bars to increase rigidity. These bars should be the same gauge as the frame of the grille.
5. If fixing lugs for the grilles are used, they should be the same gauge of steel as the outer frame of the grille.
6. Each piece of the grille should be cut to the exact length, and welded on both sides to adjoining sections of the grille.

Grille finishes

The finish used on the grille will depend on the budget available. The least expensive option is to use an oil-based paint applied by hand. A more expensive option, but saving in long-term maintenance, is to use a plastic or enamel finish. Whichever option is chosen, grilles can be finished in a variety of colours to match the interior (white is usually preferred as it matches most interiors).

Opening grilles

The grilles described so far are designed to be fixed permanently into the window recesses. Window grilles can, however, be designed to open to allow for cleaning, fire escapes, etc.

Rigid grilles which open   Rigid grilles can be designed to open or be removed. One approach could be to have grilles that can be lifted off when required, although the size and weight of the grille will determine how practical this is. Alternatively grilles can be fitted with hinges and locking devices so that they can be folded back out of the way. Again the size and weight of the grille will be a important consideration.

Collapsible window grilles   These are specifically designed and manufactured to fold away into the window recess when not in use, see Diagram 52. Although they tend to be less attractive (when closed) than decorative wrought iron grilles, in the open position they take up less space. This can be particularly useful for large windows or windows which have restrictive internal features, such as wall-mounted kitchen units on either side of the window.

Window shutters

Window shutters are a common feature in European countries and increasingly in high streets throughout the UK. In southern Europe, where shutters tend to be used as shading for buildings, they are usually an integral part of the building and are incorporated at the design stage. In the UK, shutters are mainly an add-on feature to buildings for security purposes, although there are some examples of window shutters – usually made of timber – as a feature of the old traditional shop front. In many of these buildings shutters formed part of the overall design.

For many years there has been an ongoing debate over the use and value of shutters and it would be foolish to deny that they can have significant impact on building security. They can also protect vulnerable glazing and, in particular, shopfront glazing. According to some reports, if the shopfront glass is broken to gain entry it can often cost more to replace than the value of the goods stolen. Some insurance companies may even insist that window shutters must be installed before they provide insurance cover. However, visual appearance of shutters can have a detrimental impact on an area, creating an impression that the area is under siege.

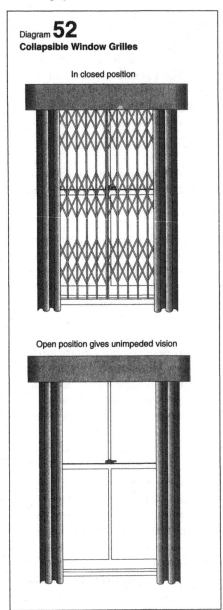

Diagram **52**
Collapsible Window Grilles

In closed position

Open position gives unimpeded vision

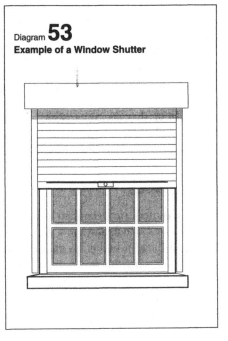

Diagram **53**
Example of a Window Shutter

The selection of shutters should therefore be given careful consideration with the aim of reducing their visual impact.

There are a number of forms of window shutter. Solid roller shutters with a bright galvanised (or similar) finish are probably the most noticeable, even when the bright surface treatment fades. They are also particularly prone to graffiti. They can, however, be finished in a variety of colours. A variation of these shutters are those which are covered with small holes or perforations, rather like a tea bag. This type of shutter can be less intrusive and, while still providing protection for the glazing, they do allow for external surveillance of the inside of the premises, increasing the likelihood of intruders being observed.

If shutters are fitted they should be made from robust materials and have securely fixed hinges, rollers and locking mechanisms. Shutters can be made from many materials, although the most common is steel (with a protective coating). uPVC shutters are also available, see Diagram 53.

Shutters can be fitted to either the inside or the outside of windows. External shutters are usually fitted to protect glazing such as shop fronts, in addition to improving overall security. However, they too can become targets for abuse and are vulnerable to tampering and corrosion. Internal shutters do not protect glazing but are less likely to be subjected to external damage, etc.

CHAPTER 6

ELECTRONIC MEANS OF SECURITY

INTRODUCTION

The traditional hardware for securing premises can be augmented by a range of electronic devices and systems. Some of these, such as alarm systems triggered by circuit breakers, have been developed primarily for individual dwellings, but the more sophisticated forms such as door entry systems have been principally designed for groups of dwellings, in particular high-rise or high-density developments.

This chapter considers the ever-increasing range of electronic security devices and systems available and their potential impact on residential security. For some time there has been a proliferation in security technology and the public's interest in them seems to go on unabated. Landlords and home owners alike are being placed under increasing pressure to adopt the latest technology in the name of crime prevention. A certain amount of caution, however, needs to be exercised in assessing their potential effectiveness. It is worth recalling, for instance, that door entry systems were oversold in the late 1970s and early 1980s as the solution to security problems in blocks of flats. Another supposed panacea, closed circuit television, has emerged more recently. These types of system are often predicated on the assumption that they can reduce personnel costs as well as crime problems. However, it is worth bearing in mind that the effectiveness of electronic security systems such as intruder alarms and surveillance schemes nearly always depends on the response and intervention capabilities of the system. Traditionally this response and intervention role has been freely performed by the police. An escalation in demand has meant an ever-increasing role for private security companies, particularly in the alarm monitoring field and patrolling of private property. It is now estimated that there are more private security guards employed in the UK than there are police officers.

Local authorities are also becoming more involved, introducing in-house security teams and staffed control centres. At a local level, the role of estate staff can be expanded to encompass security functions.

6.1 CONCIERGE SCHEMES

Many local authorities and landlords have introduced concierge or receptionist schemes in large multi-occupied blocks. Usually located in a designated area within the entrance foyer of the block, one of the basic functions of concierge or receptionist staff is to provide supervised access in to the block. Visitors to the block are expected to report to the concierge who will notify the resident of their arrival and then direct the visitor to the appropriate flat. In many schemes, staff can provide a wider service to residents such as taking deliveries and handling repairs.

Providing a concierge service covering 24 hours can be an expensive operation, requiring a minimum of six staff. An alternative approach would be to provide a partial service covering set periods, a typical example being from 8 am to 10 pm although this could vary to suit local requirements. Another approach adopted by some local authorities is to link blocks together, either physically using covered walkways to link each block to the central reception area, or by linking door entry systems and CCTV cameras to a central point as described in section 6.16. The attraction of this is that more dwellings can be included in the scheme.

When introducing a concierge scheme, it is important that the entrance foyer is designed and modified for this purpose. In practice, this usually means dividing the entrance into two distinct areas, a public area and a private area. The staffed area should be positioned so that, from the counter, staff have an unhindered view of the entrance area and the area directly outside the main entrance. Slightly raising the floor behind the counter will help facilitate this. Where there are building limitations, preference should be given to views of the entrance door and the area surrounding the lifts. This can be further enhanced by providing strategically placed convex mirrors.

Protection for reception staff is very important. The reception area must be a separate part of the foyer with access into it controlled by the staff. This does not mean, however, that the reception needs to be completely enclosed with, for example, bandit screens. A reception counter or desk which is wide and relatively high on the public side will provide a less obvious but still effective protection against assault. For children and people in wheelchairs, part of the counter will need to be at low level but this area can be protected by a screen.

Removable security screens for the whole counter should be available for use as and when required. A robust lockable shutter will also be needed to protect the reception area and equipment when staff are off duty. In high-risk areas it may be necessary to install an under-the-counter 'panic button' with a direct link to the emergency services.

Although the technological elements will vary from scheme to scheme, a typical package of equipment to aid the concierge or other housing staff could include:

- surveillance cameras at all vulnerable points, e.g. entrances and exits, lobby areas, within lift cars and in door entry panels;
- non-contact door entry system using fob or token access for residents, as described in section 6.4;

- remote opening and locking control for main door and any communal internal doors;
- two-way communication links between the control room, every resident, and all door entry points included in the scheme;
- speech transfer equipment within each lift car for use in an emergency, such as a lift braking down – and while this facility could be extended to all camera locations to enable the concierge to speak to someone in the cameras' view, it is particularly important to ensure that this does not allow staff to listen in or eavesdrop on residents' private conversations;
- video recording facilities, as described in section 6.13;
- a purpose-built ergonomically designed desk or counter for the monitoring and control equipment, to facilitate ease of use.

6.2   DOOR ENTRY SYSTEMS

Door entry systems, also often referred to as entryphones (or access control), are a way of controlling or restricting access into buildings such as offices and blocks of flats. They can also be used to control access to external areas such as car parks and communal areas on estates. They were the first extensive use of electrical technology for security purposes and, as such, were also oversold as a cure-all for crime problems in multi-occupied blocks. Many installations were abandoned, however, following numerous breakdowns, deliberate damage and the high costs of repairs.

Door entry systems have been shown to be effective, but in limited circumstances. Generally, they work best where there is a stable mature community with few children, where problems emanate from outside rather than from inside the block. Even so, a well-designed and maintained door entry system in these circumstances will not necessarily prevent unauthorised access. A more satisfactory arrangement would be to introduce on-site staff in the form of a concierge or receptionist schemes to supervise access into blocks.

A door entry system is principally a communications system, providing two-way speech (duplex speech) between communal access controlled doors and individual residents' flats. It also provides the facility for residents to open communal entrance doors remotely from their flats. The door entry system is made up of a number of interacting components which need to be compatible and robust. Failure of one component will result in the whole system failing. Experience has shown that door entry systems have a number of potentially vulnerable components. These included the entrance door locks and their cylinders, electric keeps, door closers, the entrance door and the associated joinery and glazing.

There are a number of key technical issues to consider when introducing a door entry system. In many instances, there will be an existing door entry system in place. In these cases, economies could be made in overall installation costs by modifying the existing system to digital operation. This will result in less inconvenience to residents, less disturbance to residents' decorations and fewer problems associated with gaining access to individual dwellings. However, this

Plate 12 Example of dated equipment – call panels with many buttons can be confusing to use

approach may present a number of potential problems. Many security companies, for example, are reluctant to take on work involving modifications to existing systems, claiming that any new equipment would not be compatible with what is already in place.

Alternatively, the opportunity may be taken to install complete new systems offering up-to-date technology, possibly providing improved and additional facilities. Obviously this would be more expensive initially in capital terms, but in theory savings could be made in longer term maintenance.

Another important issue to consider is where blocks are to be linked together electronically. Although this is dealt with in section 6.16, it is essential that door entry systems are compatible with the various methods of interconnecting blocks, i.e. infra-red or microwave transmissions which may need to be employed where hard wiring is impractical or uneconomical. In these circumstances it is also important that the door entry system can operate independently, should the interconnecting link become inoperative.

6.3 ENTRANCE DOORS AND FRAMES

Communal entrance doors are subjected to constant use as well as wilful damage and abuse. They have to be designed to accommodate all the necessary hardware to

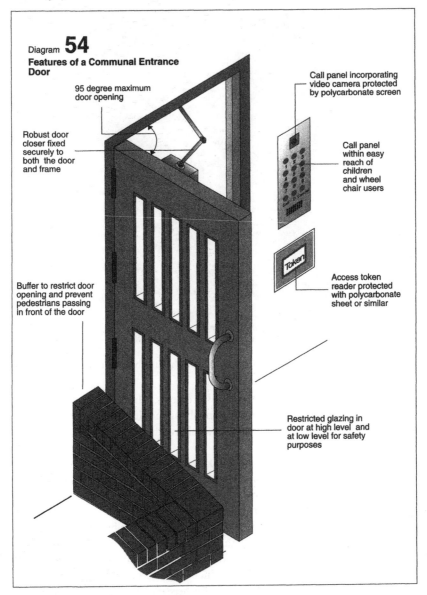

Diagram **54**

Features of a Communal Entrance Door

95 degree maximum door opening

Robust door closer fixed securely to both the door and frame

Call panel incorporating video camera protected by polycarbonate screen

Call panel within easy reach of children and wheel chair users

Buffer to restrict door opening and prevent pedestrians passing in front of the door

Access token reader protected with polycarbonate sheet or similar

Restricted glazing in door at high level and at low level for safety purposes

function as a means of restricting access. They also need to be able to withstand considerable wear and tear, yet still look attractive. These rather demanding requirements make the selection of communal entrance doors a critical factor in the design of door entry systems. Diagram 54 shows an example of a typical entrance door for a relatively low-risk environment. Many of these features are standard requirements for any communal entrance door.

Communal entrance doors should always be single leaf and open outwards. The reason for this is that the door frame and, in particular, the door stop will provide added strength to the door when it is closed, making it less vulnerable to impact force against the door from the outside. This is not the case for inward-opening and double doors which are more likely to be forced or burst in. The door opening should be at least 900 mm wide. This is the minimum distance between the door stops to allow access for pushchairs, buggies and wheelchairs.

Doors can be constructed from many designs using different materials. These include steel, aluminium, plywood or glass-reinforced plastic facings fixed on to steel, aluminium or timber (hardwood or softwood) frames. The frames are often reinforced, usually with steel inserts. Alternatively, facings can be applied to doors of solid core construction usually made up of strips or laminates of timber to form a solid raft. This method of construction produces a very strong and robust door but also one which is very heavy, placing considerable strain on the hinging mechanism and the door frame. Particular attention, therefore, needs to be given to the door frame, especially to how well it is fixed to the surrounding structure. Similar principles apply to the fixing of the hinging mechanism to the door and to the door frame.

Glazing is required in communal entrance doors as a safety measure, particularly so that people on the outside of the door can been seen before the door is opened by someone on the inside, thereby reducing the risk of accidents and unauthorised entry. Glazing should be provided at high and low level so that children can be seen as well as adults. The type and style of the glazing used for any particular location will depend on the potential risk of damage and unauthorised access. For example, in vulnerable locations, glazing should be kept to a minimum and be of polycarbonate sheet rather than glass. Where there is less of a risk, larger areas of glazing can be used. This is obviously more desirable as it allows more natural light into the foyer, which is particularly important for staffed entrances. See section 5.21 for more details on glazing materials.

Door frames

Door frames should be fixed into surrounding structures of brickwork or concrete and not blockwork. The frame should be fixed into the structure using heavy duty bolts such as expanding bolts. Four of these should be fixed in each door jamb and two into the head of the frame, if possible.

Door closers

Communal entrance doors need some form of automatic closer to ensure that the door is fully closed after use to enable the door lock to engage. This practice is fundamental to the whole door entry system, without which the system will fail. In practice, this means that a closer must have sufficient strength to shut the door but not be so strong that it prevents the frail, elderly and children from opening the door. This balance can be particularly difficult to achieve, as it is dependent on a number of factors, including the weight of the door, the need for it to open

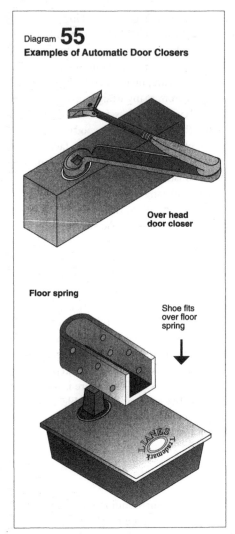

Diagram **55**

Examples of Automatic Door Closers

Over head
door closer

Floor spring

Shoe fits
over floor
spring

outwards and any prevailing winds. Down draughts within the block can also cause problems.

In very general terms there are two types of door closer that can be used on communal entrance doors: overhead closers, or floor springs, as shown in Diagram 55. Both operate in a similar manner using springs. These are tensioned when the door is opened and rely on this force to close the door after use. There are however, differing views on which are the most suitable for communal entrance doors. Overhead closers offer a number of advantages over floor springs in that they are easier to install and maintain. They also carry less weight, which means they can be finely adjusted to suit local requirements. The main drawback with overhead closers has always been their vulnerability to abuse.

Floor springs, on the other hand, are concealed within the floor and are therefore far less vulnerable to damage. They are, however, fixed close to the hanging edge which means they carry almost the full weight of the door, making fine adjustments difficult. Whichever type of closing devise is chosen, ensure that it is of suitable strength and durability for the environment in which it is to operate. Door closers should also have a back check facility which slows the rate of closing to prevent the door from slamming against the door frame. One feature they should not have is a stand-open facility which allows the door to be left in the open position. Obviously this would undermine the whole door entry system.

Door hinges

Hinges used on communal entrance doors need to be more durable than those used on individual dwellings. These doors are constantly opened and closed which would soon wear out domestic type hinges. Instead, heavy duty hinges designed specifically for this role should always be specified. Typically these have nylon washers or ball bearings incorporated into the hinge knuckle and are designed to make the hinge work more effectively. They also prolong the life of the hinge. For maintenance purposes a retaining screw is used to hold the hinge pin in place which can be removed if the washers or ball bearings need to be replaced. This screw also prevents the pin from being tampered with from the outside. A minimum of four hinges should be used on communal entrance doors.

A continuous 'piano' hinge – a common type of hinge used on many communal entrance doors – extends the full height of the door. These are generally considered more suitable for vulnerable doors because they provide better fixing arrangements between the door and door frame. They also reduce the risk of jemmying, as there are no gaps between the door and door frame that can be used for leverage. For additional anchorage, the hinge could be fixed through the door frame into the surrounding structure.

Other ironmongery

Two important items of ironmongery are fitted to communal entrance doors. On the outside of the door, a pull handle is required to open the door. This can be in the form of a simple 'D' handle, although there can be concerns that this could be used as a point of leverage to force open the door. An alternative would be a door knob (or similar). These provide less grip but can easily present serious problems for people with little upper body strength or with arthritic hands. On balance, if electromagnetic locks are used, the risk of a door being pulled open forcefully is small and 'D' handles, therefore, should be suitable.

Finally, kicking plates should be fitted to both sides of the door to provide protection from wheelchair and pushchair impact and therefore prolong the life of the door. It is easier and far less expensive to replace damaged kicking plates than it would be to replace the door. Kicking plates should be made from resilient materials such as stainless steel and be fixed securely onto the face of the door.

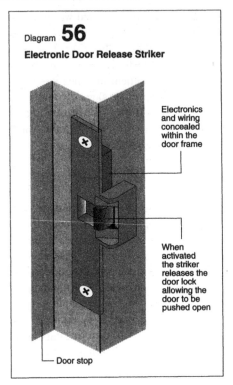

Diagram **56**

Electronic Door Release Striker

Electronics and wiring concealed within the door frame

When activated the striker releases the door lock allowing the door to be pushed open

Door stop

6.4 LOCKING SYSTEMS

The standard method of releasing communal entrance doors in the past has been by the operation of electrically controlled lock keeps or strikers, as shown in Diagram 56. Used in conjunction with mortice locks, these are mechanical devices designed with a spindle which allows the striker (or, keep) to move and release the door lock, when activated from one of the residents' handsets. They are particularly vulnerable, however, and are constantly subject to misuse and vandalism. If force is applied to the door, there is a tendency for the spindle to buckle, rendering the striker inoperative. Often the only course of remedial action is to replace the striker. This type of locking arrangement is really only suitable for controlling access to a small number of dwellings, up to about six.

Magnetic locks offer an alternative for communal entrance doors serving many dwellings. Shear magnetic locks are installed to the leading edge of the door and provide a mechanical lip and groove connection, in addition to the magnetic hold, to enhance security (see Diagram 57). However, this mechanical connection relies on very precise alignment of doors and may present problems of misalignment due to general wear and tear.

Face magnetic locks, as shown in Diagram 57, are installed to the internal face of the door and door jamb and give face-to-face contact. They are used more often

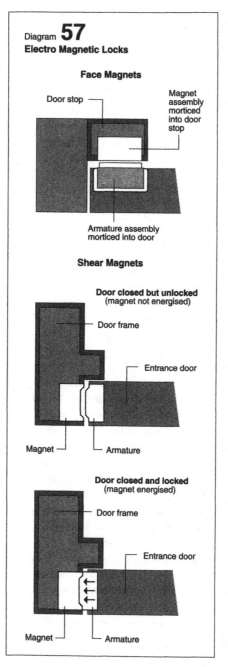

Diagram **57**
Electro Magnetic Locks

Face Magnets

Door stop

Magnet assembly morticed into door stop

Armature assembly morticed into door

Shear Magnets

Door closed but unlocked
(magnet not energised)

Door frame

Entrance door

Magnet

Armature

Door closed and locked
(magnet energised)

Door frame

Entrance door

Magnet

Armature

because they are more tolerant to minor misalignment of doors and less prone to external abuse when in the locked position.

The number of locks required will depend upon the quality of the door frame (all doors must be rigid in construction and not subject to warp or distortion). In normal circumstances, a single centre-fixed lock may suffice. However, two locks – one to the top and one to the bottom of the door – will provide a high degree of security. Magnetic locks are exposed when the doors are open and can be rendered ineffective by masking with insulation tape. This, however, is far easier to rectify than would be the case if a conventional lock and electric striker were damaged and needed to be replaced.

6.5   RESIDENT ACCESS

Entry for residents through communal entrance doors can be provided in a number of ways. The conventional means of lock and key is the least expensive in capital terms but can also be the least effective, particularly where a large number of keys are involved. This method is vulnerable to abuse by physical force, manipulation and unauthorised key cutting and, again, is only suitable for small numbers of dwellings.

A number of electronic-based alternative systems are available, ranging from magnetic cards, optically encoded keys and electronic fobs or tokens. Most of these systems work on some form of magnetic code, similar to those found on credit cards and the like. Each token or fob has its own unique code which, providing it is recognised by the system, will activate and release the lock and allow for access. It is claimed by manufacturers that there are millions of codes available. Some card systems require the card to be inserted into a slot or swiped along a groove to activate the system. These systems may work well in closely controlled or managed environments, but in public areas, such as entrances to blocks of flats, the slots or grooves can be particularly vulnerable to abuse and vandalism.

The system considered most suitable for housing and, in particular, multi-storey blocks is based on an electronic token. This system has no moving parts, vulnerable slots or key holes. Instead, to operate, the residents simply present their token to a reader which is usually positioned close to the locking edge of the door, and, if recognised, the door lock is automatically released.

One of the main advantages of this type of system is that lost or stolen tokens can be programmed out of the system so that they cannot be used for unauthorised access. Some systems are managed by the installer of the system. In practice this means that they need to be notified of lost tokens, which can cause delays in having them deleted from the system. A better arrangement offered by other systems is where the programming can be done in-house without having to rely on a third party. This is a relatively simple operation involving a personal computer (the system controller) and a familar operating system – Windows 3.1 or Windows 95.

Another advantage of the electronic token is that it does not require accurate positioning in order to operate. This is particularly useful for residents who are visually impaired or whose disabilities make using a conventional key difficult.

Plate 13 Resident using non–contact access token

These systems offer a number of facilities including: access control for hundreds of doors, monitoring pedestrian traffic, monitoring door status (open or closed), monitoring alarm inputs and networking of several schemes to a central point capable of being supervised by a concierge or receptionist.

The system controller must be sited in a secure cabinet located within a room accessible only to staff. To be effective in controlling the door access system, it should be capable of providing the following facilities:

- fully programmable facilities, preferably by in-house staff;
- token recognition and transaction storage;
- validation and voiding of tokens;
- door status monitoring;
- auto–timed door lock/unlock periods;
- integral door release;
- option for obtaining a print-out of all transactions going through the system, so that movements and use of tokens can be monitored – but before taking up this option it will be necessary to check whether the acquisition of such data contravenes any laws or codes of practice on data protection and civil liberties;
- standby battery supply to maintain systems for a minimum of 3 hours in the event of a mains failure.

When selecting door access tokens, ensure that they are:

- fully encapsulated units without moving parts or batteries;
- made of robust, high-impact plastic;

- programmed with a unique code during manufacture;
- of a suitable shape and size to enable them to be easily handled and attached to a key ring;
- have a long life expectancy with a minimum five-year guarantee.

Token readers should ideally be recess mounted next to the entrance door. The back box (containing the electronics, etc.) can then be concealed within the structure of the main entrance to prevent tampering or damage. A non-metallic vandal-resistant polycarbonate plaque should cover the reader unit and be engraved in the shape of the outline of the token or a key with the word 'here' in the centre to illustrate where the token should be presented.

All communal entrance doors should be fitted with an exit button or pad on the inside to facilitate egress. This should be situated on the door jamb adjacent to the opening side of the door and fixed at a height that permits its use by children and wheelchair users. This button should be of the same vandal-resistant design and quality as the main door entry panel.

There is some debate over the number of tokens that should be issued. Some housing authorities issue two tokens per dwelling, whereas others only provide a token for each person named in the rent book. However, it may be necessary to provide extra tokens where there are older children living at home, family visiting elderly relatives, etc. In such cases extra tokens can be issued, although it could be argued that the more that are issued the more the system becomes liable to abuse. However, door entry systems are not infallible and even very sophisticated systems can be overcome.

6.6 VISITOR ACCESS

Call panels are used by visitors to contact residents in order to gain access to the block. They are the public part of the communications system, usually situated next to the communal entrance door and, as such, are vulnerable to misuse and vandalism. Careful selection of the call panel is therefore rquired.

Simple, low-cost control panels are available for small schemes; the example shown in Diagram 58 is suitable for three flats. This is surface mounted, not particularly robust and therefore unsuitable for vulnerable locations.

A stronger alternative is shown in Diagram 59. This type of call panel should be recessed slightly to protect vulnerable edges, and fixed with concealed or anti-tamper fixings. Facia plates, while needing to be attractive in appearance, should be made of a tamper-proof material such as stainless steel with fully flush push button controls. Panels should be surrounded with robust materials such as masonry or concrete. Timber (soft or hardwood) and metal trims, etc., should be avoided as they can be easily damaged, exposing the vulnerable edges of the facia.

Numerical display windows should be protected with 6 mm clear polycarbonate glazing. Call buttons and all polycarbonate glazing fixings should be strong enough to withstand heavy impacts without adverse effects. Any speaker microphones incorporated within the panel should be protected with heavy duty gauze fixed

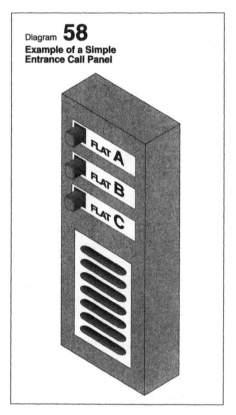

Diagram **58**

**Example of a Simple
Entrance Call Panel**

behind the outer grill in order to protect them from misuse and insertion of foreign objects and liquids.

All apertures, windows, buttons and facia panels should be adequately sealed in order to prevent moisture penetration.

Some types of call panel also incorporate surveillance cameras. This will be dealt with in section 6.10 (CCTV). The camera housing within the panel should have enough space to allow for adjustment of the camera positions to ensure maximum coverage of the entrance area. The control buttons on the panel should include numerical dialling (numbered 0 to 9), 'call' and 'cancel' buttons and a button to summon the concierge (where applicable). The complete call panel should be incorporated into a robust steel back box which totally encloses the electronic components and cameras to afford protection from misuse and vandalism. This back box should be an integral part of the entrance construction and should be incorporated into the building structure. Contractors will need to liaise with builders, joiners and door manufacturers to ensure that all dimensions and fixings are compatible.

Full operating instructions should be engraved on the facia panel or on an adjacent stainless steel plate and should be easy to understand and follow. This may mean that instructions need to be provided in more than just English.

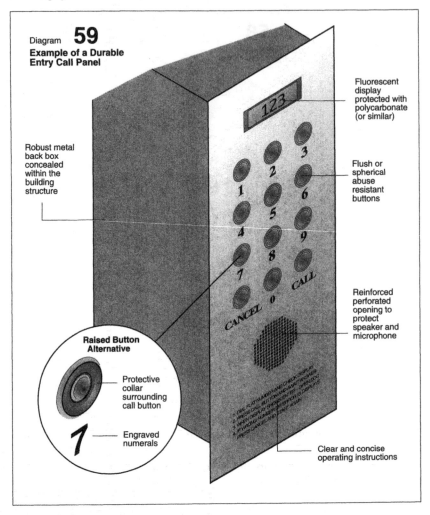

Diagram **59**

Example of a Durable Entry Call Panel

Fluorescent display protected with polycarbonate (or similar)

Robust metal back box concealed within the building structure

Flush or spherical abuse resistant buttons

Reinforced perforated opening to protect speaker and microphone

Raised Button Alternative

Protective collar surrounding call button

Engraved numerals

Clear and concise operating instructions

6.7   RESIDENTS' HANDSETS

At the other end of the communications system, within each flat are the residents' handsets. These enable residents to speak to callers (and concierge staff where applicable) and are available in various designs, some in the form of a small box fixed to the wall while others resemble telephone handsets. Some also incorporate small video screens so that residents can see as well as speak to callers. (See Types of CCTV scheme, page 117.)

Handsets are often located close to the entrance door of the dwelling, although residents should be consulted about the most suitable position. They may, for instance, prefer them to be located within the living room.

Handsets should have a number of features, including the following:

- a call tone with adjustable volume control – and a different call tone to inform the resident that the handset has been left off the hook;
- a privacy or nuisance switch so that the resident can disconnect from the door entry system for a given period of time – which can vary from as little as 30 minutes up to 12 hours, to suit the resident or can reset the system by re-pressing the privacy switch;
- the entrance door release button in each flat should only become active when a dwelling is called up from the communal entrance door call panel; this will prevent any missuse of the system or the communal entrance door from being opened accidentally;
- the facility for residents to call-up the concierge from their dwelling and vice-versa;
- calls between each resident and the entrance call panel (and concierge where applicable) should be private;
- adjustable brightness, colour and contrast controls for handsets which include video screens;
- the facility to connect aids for residents with impaired hearing.

Where two or more communal entrance doors are controlled by the door entry system, a door release button should be provided for each entrance door. It should not be possible to open all the entrance doors by pressing a single button. If a handset is damaged it should not affect the function or operation of the rest of the system.

6.8 TRADES AND DELIVERIES

Access provision for the delivery of post and milk needs to be considered during the planning stage of an access control system. Where there is to be a staffed reception such as a concierge, this should not be a problem providing staff are on duty during the daytime when deliveries are made. Problems could emerge where schemes are only staffed part time (not covering the morning period) or there is no staffed reception. In these circumstances, there is need for alternative arrangements.

One option for schemes using access tokens is to issue tokens to authorised regular callers which they can then use to gain access. The advantage of this is that should a token be lost or stolen, it can be programmed out of the system and therefore no longer be used to gain access. Another option is that digital access systems usually have the facility for regular callers to be issued with individual PIN (personal identity number) codes which are used to gain access. It may be necessary to change these codes periodically to ensure that they do not become widely known and open to abuse.

An alternative approach would be to provide a trades button on the call panel which allows the door to be opened during a set period of time, usually in the morning for an hour or two. The main drawback is that this button can be used by anyone to gain access.

There should be no special arrangements for irregular deliveries. These should always be treated as any other visitor to the block. It is not uncommon, however, for schemes involving staffed receptions to take deliveries on behalf of residents who are not available (at work, out shopping, etc.). This is often an *ad-hoc* arrangement, although guidelines should drawn up in consultation with residents, clearly defining what staff can and cannot accept on behalf of residents.

6.9   EMERGENCY ACCESS

Responsible landlords always consult with the local fire service over the means of access to a block of flats in the event of an emergency. The provision of override switches is probably the most common approach to this. There are a number available using different types of access key. These are usually positioned close to the entrance door but out of reach to prevent tampering and manipulation.

Whenever access control schemes are introduced or up-graded any changes made which impact on emergency access should be agreed in advance with the local fire service. It is also important that the approach adopted is consistent where landlords are responsible for a number of access control systems.

6.10   CLOSED CIRCUIT TELEVISION (CCTV)

Over the last few years there has been a great deal of positive publicity about the effects of CCTV surveillance on crime and anti-social behaviour. Apart from the long-running television programmes which use images from cameras to help police with their enquiries, there are weekly programmes totally devoted to showing the latest CCTV footage of high-speed car chases involving the police in pursuit of so-called joyriders or speeding motorist on the motorway network. Most of this attention has been concentrated around public areas such as town centres, public highways and public buildings. This is not to say that CCTV is not useful in a housing environment. CCTV can and does have many applications. However, it is important to recognise that CCTV schemes that have had some success have been introduced in particular circumstances. Most significantly, many schemes have been introduced as part of an overall management strategy for an area, involving close collaboration between all interested parties. In effect they have been introduced as aids to management and in tandem with other measures, rather than as free-standing security initiatives.

When devising a CCTV scheme it is important to be clear about what the scheme is meant to achieve. There have been well-publicised schemes recently in which images from video recordings have been used to locate and take proceedings against offenders. Invariably, the capture of such images will be of more use in tracking down offenders where they are already known to the police. Images of people not already known to the police may be of little use unless offenders are caught in the act. If the main target of a CCTV scheme is serious offenders such as those involved in violent theft from businesses, then a scheme which records images but has no constant monitoring may be regarded as adequate. However, if the scheme is intended to be used to deter lesser offending and anti-social behaviour, and many of those involved are unlikely to be readily recognisable to the police, then recorded images may not be enough and staff may need to be deployed to monitor and respond to the images on the screens. It is our experience that young people in particular soon realise that, where the presence of a camera does not lead to any kind of immediate response, the chances of their being caught are going to be slim.

Types of CCTV scheme

At its most basic level an inexpensive camera can be linked to small monitor or even a spare channel on a domestic television set so that a householder can identify callers at the front door. This appproach can be extended in a block of flats to enable all flat dwellers to identify who is standing at the main entrance to their block and to monitor internal and external communal areas. The one drawback of this type of system is that it can be used by miscreants living in the block to monitor the movements of other householders, to enable them to judge a suitable time for a burglary. The other end of the scale involves the extensive use of cameras monitoring whole blocks of flats and estates linked back to dedicated control rooms which are monitored by staff. The reduction in the size of cameras has also meant that they can be positioned virtually anywhere, even in confined spaces such as inside lift cars.

Surveillance schemes can be used for specific purposes. For example, some local authorities have used covert cameras to obtain evidence of victimisation and racial abuse against individual residents. These have been temporary arrangements and the cameras and associated equipment removed once the problem has been resolved. Strict guidelines and procedures are always necessary for schemes involving the use of covert cameras and these safeguards should be made public.

6.11 CAMERAS

One of the first things to be considered when introducing a CCTV surveillance scheme is whether the area which is going to be monitored can be adequately covered by a number of fixed cameras or by a single movable camera.

Fixed cameras, are rigidly mounted and targeted at single specific areas. Fixed cameras are generally more suitable for monitoring internal areas such as entrance foyers and inside lift cars, although they can be used externally, for example, to monitor entrance doors or approaches to blocks. Movable cameras, on the other hand, usually referred to as PTZ cameras (pan, tilt and zoom) can scan a much larger area and allow an operator to select precisely an area to be monitored. The motorised zoom lens then enables the operator to zoom in to enlarge a particular part of the area under surveillance in much more detail. These cameras are far more expensive than fixed cameras, both in capital terms and ongoing maintenance costs.

However, it may be more cost effective to have one PTZ camera rather than a number of fixed cameras to monitor an area. For this reason they tend to be used to monitor large open areas such as car-parking areas or street scenes in town centres. PTZ cameras can scan an area automatically but, to be effective, they really need to be monitored constantly by staff.

Another consideration is whether to use a monochrome or a colour system. Monochrome has the advantage of being less expensive than colour, although the price difference is rapidly narrowing. Another advantage is that it can operate at much lower lighting levels. This should not really be a consideration, however, as the provision of comfortable external lighting levels for people – a priority for community safety – should also be suitable for colour surveillance systems.

Colour, however, has a number of advantages. It enables the operator to identify people easily through the colours of their clothes, hair, etc. Watching colour images is also more pleasing for staff than watching drab monochrome and may even encourage better use of the system.

Lenses are the eyes of a surveillance system, having either a standard or wide-angle view, and should be selected to suit the environment in which they are to operate. They will, to a large extent, determine the quality of the image displayed on the monitor.

Camera housings and wall-mounting brackets should be made of robust vandal-resistant materials and the design should allow for them to be fixed securely to the wall or ceiling. All fixings used should be concealed or tamper-proof.

6.12 MONITORING EQUIPMENT

The selection of equipment required to monitor the images from cameras will depend, to a large extent, on the number of cameras involved. Obviously, if only a single camera is being used, for example, to monitor a secluded entrance door, then this can be connected to a small monitor to constantly display the image. This approach could be used to monitor a number of cameras but this would be expensive as each camera would require its own monitor. Switchers have been developed to do away with most of these monitors. A sequential switcher is probably the most popular switcher. In normal operation, these display one picture at a time from each camera in succession. The length of time each camera image is displayed can be adjusted as well as the order in which the images are displayed. They also should have the manual facility to select and stay with an image from a chosen camera. As a general rule, sequential switchers are only suitable where a small number of cameras are in operation.

A more sophisticated switcher – a matrix switcher – has been designed for larger schemes involving many more cameras and monitors. However, these have been superseded, to a large extent, by multiplexers. These can constantly display up to 16 camera images onto a single monitor screen at one time as well as enabling all of these images to be recorded onto a single video recorder. A separate monitor (or spot monitor) can be provided to display the images from any chosen camera in full

Plate 14 A typical multiplexer

screen size to allow clarity of view. A multiplexer can also operate in a similar way to a sequential switcher.

Monitors should be designed specifically for use with CCTV systems. Although similar to televisions they have higher lines of resolution which produce better picture quality. When selecting monitors, particular attention should be paid to the size of the screen. Sizes are given in inches measured diagonally across the actual screen. As a general rule, always select the largest size monitor that funds will allow and never less than 12 inches for full display or 17 inches where multiple camera positions are being monitored.

6.13   VIDEO TRANSMISSION

The purpose of video transmission is to relay the video signal from the camera back to the monitor. There are a number of methods that are available such as coaxial cable, telephone lines, fibre optic, infra-red and microwave. Each of these methods has its own advantages and disadvantages and for this reason it is not uncommon to find more than one method used in a CCTV surveillance system.

There are a number of factors that will influence the selection of the video transmission method, such as the distances between the cameras and the monitors, the environment, the layout of the site and the cost. In addition, each method will be subject to various forms of interference. A well-designed CCTV scheme will take account of these limitations to ensure that they are kept to an absolute minimum.

Coaxial cable is by far the most common method of linking cameras to monitors over short distances and is the most likely method to be used on estates and blocks of flats. The cable is shielded to screen the transmission signals to reduce any external interference such as electric cables. This type of transmission is not particularly suitable for use over long distances due to video signal loss, although signal boosters can be introduced. Instead, fibre optic, infra-red and microwave systems can be used. These are described in more detail in section 6.16.

6.14 RECORDING CAMERA IMAGES

The need for recording images from CCTV cameras will depend, to a large extent, on what the surveillance system is used for and how it is managed. The various methods of recording images from CCTV cameras are given below.

Real-time recording (standard recorders)

These VHS video recorders are similar to domestic recorders but are more durable (and expensive) and designed for far greater usage. They capture images from cameras as they happen.

Time-lapse recording (time-lapse recorders)

This type of recorder is used in the majority of CCTV systems and records frames with gaps between them. Typically this means that there are around one-third fewer frames per second than if recorded in real time, although this number can be reduced even further. The advantage of this is that it extends the length of recording time of a single video tape by as much as 24 hours. Time-lapse recorders can be switched to real-time recording by an operator if an incident is observed. When tapes recorded in time-lapse mode are played back, any movement will appear jerky.

Recording facilities on standby mode

Standby recording provides the facility to record only as and when required, as well as continuous recording when necessary. This keeps capital, maintenance and tape management costs to a minimum. More sophisticated systems are available which trigger the video recorder automatically by detecting movement within the field of the camera (see section 6.17).

Hard copies or stills

These are provided by a video printer which is an item of equipment that can be added to surveillance systems. They provide instant hard copies or photographs from the camera images.

Continuous recording

The provision of 24-hour continuous recording requires a considerable library of tapes and management time to change and monitor tapes. It involves a high capital outlay and the continuous running of recorders means increased service and maintenance costs. When considering the need for 24-hour continuous recording, it must be recognised that the areas under surveillance are only a small proportion of the total areas of an estate or block exposed to abuse and vandalism. Seen in this context, continuous 24-hour recording may not seem as necessary, particularly for schemes which are staffed. One option may be to only use recording facilities for schemes which are staffed part time or not at all. Video tapes can then be viewed at a later date, if an incident is reported. Continuous recording can be used in conjunction with staff to prevent the system from being misused (i.e. to prevent close-up views through dwelling windows), a potential problem especially where pan, tilt and zoom cameras are in operation.

6.15 CODE OF PRACTICE

In order to ensure that a CCTV surveillance scheme, which involves a number of dwellings such as as an estate or block of flats, runs smoothly and in accordance with the criteria which was agreed during the planning of the scheme, a code of practice needs to be drawn up between all interested parties. Usually this involves the residents through their association or similar body, the landlord of the dwellings such as the local authority, the police and, depending on the make-up of the area, other groups such as local businesses. Although the code of practice should be tailored to suit local circumstances, it should clearly define the complete operation of the CCTV scheme covering issues such as access, storage and confidentiality of recorded information (video tapes). The Local Government Information Unit has produced a guide *A Watching Brief: A Code of Practice for CCTV* (1996), which provides detailed guidance on drawing up codes of practice.

6.16 LINKING SYSTEMS

Over recent years technology has advanced to allow door entry and surveillance systems to be linked to a central control point for the purposes of monitoring and supervision. In practice, it is now possible to manage entrance doors, lifts and surveillance cameras from a number of different sites using staff at a single location. This approach, often termed 'dispersed concierge', has many attractions to landlords, not least the opportunity to provide supervision and intervention at a number of locations using a relatively small number of staff. How effective these schemes are will inevitably depend on local circumstances. A five-year evaluation of these schemes has been carried out by the Safe Neighbourhoods Unit on behalf of the Department of the Environment and the findings are published in the report *High Hopes: Controlled Entry and Similar Schemes for High Rise Blocks* (1997), HMSO.

In terms of the technology involved, there are two basic ways door entry and surveillance systems can be linked.

Hard Wire

This involves physically linking each block to the central point using cables which are usually laid in underground ducts specifically for this purpose. Depending upon the distances involved, this can be expensive, especially if roads and hard landscapes need to be disturbed and reinstated after the ducts have been laid. However, once installed, hardwiring is not prone to vandalism and requires little or no maintenance. Fibre optic is considered to be the most suitable for long-distance transmission cabling as it is not affected by interference such as electric cabling (which can run in the same duct).

Where site details allow, it may be possible to utilise existing links such as telecommunication lines. There may, however, be limitations on the type of equipment that can be connected to these lines. There is also the cost of line rental which can vary from area to area. These costs need to be compared with other options of linking blocks, such as dedicated lines as described above or free air transmission described below. Ongoing maintenance costs of equipment should be included in this equation.

Free air transmission

This method involves transmitting data between blocks of housing and the central point through the air. Usually microwave transmitters are used for this purpose although it is possible to use infra-red frequencies. This alternative to hard wire is only suitable where there are clear sight lines between each transmitter. Relays can be used to go around or over any obstructions but this will increase the installation and maintenance costs.

One of the main advantages of these systems is that they can be installed quickly with the minimum of disruption and, depending on the distances involved, at a lower cost than ducting. However, the equipment needs to be exposed, usually mounted on the roofs of tall blocks, which makes it prone to vandalism.

Free air transmitters are also affected by weather conditions. They can, for example, be adversely affected by snow, rain and fog to the point that the system becomes inoperative. Wind can be a problem for particularly exposed sites blowing equipment out of alignment and causing the system to fail. Some systems are more prone to climatic problems than others, microwave being considered least affected by the elements. A licence may be required to operate a microwave system.

6.17   RECENT DEVELOPMENTS

There have been a number of recent developments in security technology, such as systems designed to protect other security equipment. This concept is not new – cameras have been used before to monitor door entry panels, but this has been taken one stage further. Alarm systems are available principally to protect surveillance cameras. They are designed to be able to sense when a camera is being tampered with. The system then alerts staff and activates other cameras so that they can monitor what is happening as well as switching on video recording facilities. It is difficult to assess how effective these systems would be although it is likely that much will depend on how quickly staff can respond to an incident.

Another development has been video motion dectors (VMD). These devices are designed to activate cameras and recording equipment or set off an alarm if movement is detected in the area under surveillance. Clearly this system is not suitable for general use in blocks of flats or estates but may have specific applications where the general public does not have access. For example, reception areas when staff are off duty or vulnerable lift motor rooms in blocks of flats. They could also be used to monitor emergency exits and community facilities when not in use. It is worth noting, however, that the cost of installation may far outweigh any benefits this type of system could offer, with the possible exception of control rooms which house valuable equipment.

6.18   ALARM SYSTEMS

Although estimates suggest that over 90 per cent of alarm soundings from premises are false, they are still very popular for deterring burglars. For domestic alarm systems that are not linked to a central control point it is probably more important that they sound loudly inside the building, to panic the intruder, rather than outside where they may be ignored (at least in the first few minutes) by neighbours immunised by too many false soundings. If an alarm system is installed, it is essential that potential intruders are made aware of this before they attempt to gain entry, either by a prominently located sounder box or by warning stickers placed strategically on doors and windows (don't forget the back!).

Although some alarm sounders (whether stand-alone or linked to central control points) are set off by users pressing a 'panic button', most alarm systems are designed to be triggered inadvertently by the intruder. The three most common triggering devices are circuit breakers, pressure pads and movement detectors. Circuit breakers can be fitted to doors and windows. When the doors and windows are closed, two small contacts (one on the frame and one on the opening section) are in sufficient proximity to complete an electrical circuit. When the door or window is opened, the circuit is broken and the alarm sounds. Circuit breakers can be wired up in series so that all the ground floor openings in a building can be monitored by one circuit.

Pressure pads consist of thin sheets which can be laid under mats or carpets. When stepped on they trigger the alarm circuit. Pressure pads are typically fitted under the carpet at the top or bottom of stairs.

Movement detectors can trigger alarms by sensing vibrations, pressure changes or the infra-red body heat transmitted by an intruder. Vibration detectors are commonly used for car alarms but can be attached to sheds and outbuildings where they will be triggered by attempts at forced entry. Passive infra-red detectors are sensitised by the heat of a body passing through their invisible beam area. They can be mounted in rooms and hallways and are also commonly used to trigger external security lights. Their sweep and sensitivity have to be carefully adjusted so that they are not constantly triggered by our feline and canine friends!

Movement detectors are particularly useful in rooms with large areas of external glazing such as patio doors which an intruder may smash in order to gain entry. In these circumstances, contact breakers would not trigger the alarm as the doors would still be shut, but a movement detector within the room would.

Until recently, alarm systems for individual dwellings have been hard wired, linking each contact breaker, movement detector, etc., back to the control box using cable. This can create unsightly cable runs or trunking around dwellings. Over the last few years, wireless alarm systems have become increasingly available to overcome this problem. These rely on radio signals rather than cable to transmit data from movement detectors to the control box, and are also easier and less expensive to install than their wired counterparts.

Portable alarms

Many landlords suffer from vandalism and damage to dwellings which are vacant, undergoing refurbishment, between tenancies or simply waiting for a new tenant to move in. Conventional methods of securing these dwellings is by boarding them up, which is not only unsightly but clearly indicates that a dwelling is empty.

Portable alarms can offer an alternative approach or, in vulnerable locations, provide an additional security measure. These devices are triggered by vibration or by movement sensors which activates a high-volume alarm. Some devices can also transmit an alarm signal either by radio or through the telephone network to a central response point. As with any alarm, its effectiveness will depend, to a large extent, on the response to it. Portable alarms are sometimes used as a temporary measure in the homes of people at risk of repeated victimisation, for example, racially motivated attacks.

Personal alarms

These devices are small enough to be carried and are designed to be used by individuals if they are confronted or are in fear of their own personal safety. Personal alarms emit a high-pitched sound (shrill) when activated to frighten off would-be perpetrators and to raise the alarm. Many local authorities and other organisations issue personal alarms to staff who may feel vulnerable because they work in the evenings or on their own. They should always be issued to staff in addition to, rather than instead of, other measures and working practices designed to ensure staff safety.

6.19 TIMED SWITCHING DEVICES

These are the simplest electrical security devices to install and are effective insofar as burglars and thieves prefer to enter unoccupied premises and to leave unseen. At their most basic, timers switch internal dwelling lights on and off to give the impression of occupancy. More sophisticated devices can be programmed to switch on and off randomly or according to darkness. It is possible to use timers to trigger a wide range of appliances such as radios, televisions, or even motors which draw curtains. A man in Milton Keynes has taken all this even further – if he is held up at the office or on a business trip to another part of the country, he can send a telephone signal, via modems, which will trigger an electronic controller to draw his front room curtains and switch on the lights in his home. With the rapid increase in home computers, it will not be too long before this kind of integration will be commonplace. Software programs will become widely available which will be able to control lighting in every room, close and open curtains, regulate heating and, to the outside world, create the impression that not only is the dwelling occupied, but people are actually moving around inside, watching television and even going to bed!

CHAPTER 7

PLANNING AND IMPLEMENTATION PROCESS

This chapter is mainly concerned with the process of up-grading existing buildings and developments. However, many of the topics covered are also relevant to new build.

7.1 PLANNING AND PROGRAMMING THE WORK

Once the specific problems of a neighbourhood or housing development have been identified and the improvement measures decided upon, it is essential that any programme of work is carefully planned in a series of 'self-contained' phases. There are two reasons for this. Firstly, many housing authorities face uncertainties over future funding, and it is important that any phase of work undertaken is independently effective. Secondly, any initiatives depend to a great extent on the good will and co-operation of residents. If a lengthy programme seems to be providing no more than continual disruption, the goodwill of residents will quickly evaporate.

It may be necessary to make modifications to the design during the construction process, and it is important to be prepared to make any essential changes in response to feedback from residents. If the work has been planned in phases it is important to recognise the effects of any modifications on future stages of the work. The need for modifications may only become apparent after a trial period of use, or as circumstances in the neighbourhood change. Again, consultation with residents is the most effective way to identify areas where changes need to be made.

7.2 SPECIFICATION AND CONTRACTS

The specification of materials needs to be given careful consideration and the materials should be robust, easily maintained and easily replaced. Design features that are in areas of public use obviously need to be especially robust, and will quickly fail if they have been poorly constructed or incorrectly installed. The same is true of any target-hardening measures fitted to individual dwellings.

Many security improvement programmes are complicated by the involvement of several different types of contractor. For example, the installation of a door entry system (including CCTV) involves three types of contractor: building contractor, to make building alterations, construction of the entrance foyers, etc., including all necessary joinery; electrical contractor, to provide the necessary power supplies to the cameras and other electrical equipment, and to upgrade the lighting where appropriate; and electronic contractor, for the supply and installation of CCTV, entry systems, video and other equipment. Ideally, one contractor should be responsible for the installation of the whole system. Builders tend to be reluctant to do this, although they may be considered the most appropriate because of their experience of managing subcontractors.

7.3 SECURITY AND SAFETY DURING IMPLEMENTATION

Security problems may well increase during the construction or improvement programme period. Apart from the obvious opportunities for theft of materials and tools, strangers will be less conspicuous when building workers have access to an estate as a whole, and also to individual dwellings.

If scaffolding is erected around buildings which remain occupied during refurbishment, extra precautions should be taken. The occupiers should be advised of their increased vulnerability and, where appropriate, should be offered free installation of window locks and other security aids. Outside working hours the contractor should ensure that unauthorised access is prevented by removing or chaining ladders, fixing sheet metal and overhangs at key points and installing security lighting.

Building sites are often magnets for bored youngsters and it is therefore most important to secure equipment and materials that can be misused. Rubble and inflammable waste should be removed as quickly as possible from any areas to which the public have access. Builders' materials, equipment and waste may be used as aids to crime and vandalism if they are not properly secured. Rubble can provide ammunition to break windows. Paint and cement can become children's 'play' materials. Timber stock and waste can be used for starting fires.

If outside contractors are used, the provision of on-site security and a secure contractors' compound should be part of the terms of contract. If a council or housing association use their own workforce, security requirements should be considered as part of the plan of work. The co-operation of residents is essential.

Contractors need to be reminded of security works and equipment, and should sequence the installation process with security in mind. There have been several instances of CCTV cameras being stolen before they were linked up to the monitoring area.

For major works requiring a site hut and compound, security measures should include a 2.4 m perimeter fence topped with at least one strand of barbed wire and with view slots at eye level. The site hut should be elevated for a view over the site and for load inspection. There should be a 915 mm minimum gap between the site fence and site hut. The door and shell of site hut should be attack resistant (e.g. 18 mm ply), and the roof should be as strong as the walls. There needs to be laminated glass or polycarbonate windows in the site hut or, failing that, lift-off or hinged grills or panels locking over the windows. There should be deadlocks to doors on site huts.

Security lighting needs to be elevated for night 'natural' or 'formal' surveillance. Plant, vehicles, pumps, etc., need to be immobilised at night with keys or vital parts secured in a key box in the site hut. All tools should be kept in secure lockers and loose components secured in 'immovable' amounts and/or to ground stakes. Plant and equipment needs to be clearly and permanently marked and secured if too large to be put away.

7.4 MONITORING CONSTRUCTION AND INSTALLATION

It has sometimes been assumed that, once an estate improvement plan is drawn up and approved, the implementation phase will happen automatically and that the responsibility can be passed on to the contractors. Nothing is further from the truth. Residents may have spent years lobbying for improvements; now they often have to experience months of discomfort, inconvenience and frustration. Whether the improvements involve security fittings, new fencing or landscaping, building facelifts or other major alterations, good project management is essential to ensure that the works are undertaken properly and correctly phased, that problems are identified at an early stage and that the completed works are of a high quality.

The strongest of locks will not act as a deterrent if they are incorrectly fitted or if the door and window frames are rotten or insubstantial.

Throughout the construction process it is essential to carry out checks to ensure that the contractors are actually using the specified materials, and that any new features or security measures are correctly and securely installed.

7.5 IMPLEMENTING IMPROVEMENTS

A project team should:

- include technical staff able to liaise directly with residents;
- provide efficient and properly supervised site management;
- work closely with local housing management staff and residents' representatives;
- co-ordinate the input of various departments and contractors efficiently; and
- ensure that the contractor keeps the site secure and safe at all times.

Where there is a local estate officer, an estate management team may oversee the improvement project. A project technical officer may be appointed and an estate working party set up, which might include the estate management staff, a surveyor, an architect, a contractor/tenant liaison officer, a caretaker and residents' representatives. This working party can plan and oversee the improvement works.

7.6 INSTRUCTIONS TO USERS AND EVALUATION

The dwellings or communal facilities must be handed over quickly to those who will be responsible for them. On moving in, residents must be made aware of how any security installations, from window locks to alarm systems, work; one reason why many security measures fail is simply because they are not used correctly, or not used at all.

To ensure that the full benefits of a concierge and door entry system are used and understood, three sets of operating instructions are required.

- *Residents' instructions*: Simple clear step by step instructions for operating the system including what to do in the event of a problem. Written instructions may need to be produced in different languages, including braille. Residents may also need to be shown how to use the system.
- *Concierge instructions*: A comprehensive detailed set of operating instructions is required in the form of a step-by-step explanation of the various controls and functions and the order in which the operations should be performed. This should also include instructions on dealing with problems and the procedure for having repairs carried out.
- *Technical manuals*: These include all installation, maintenance and service manuals and all necessary information for the purpose of maintaining the system.

7.7 EVALUATION

Once a system is up and running it is important that it is closely monitored. Accurate records of problems need to be kept, detailing the problems, the causes, the remedial work necessary and the cost involved. This will highlight difficulties with the system and the performance of individual components. It should also provide valuable information on where additional or alternative measures for improving safety and security are required.

In terms of effective redesign, the hand-over after completion of works should not be the end of the story. Where residents have been moved out temporarily, it is only after they have moved back into their homes and resumed their normal daily life that it is possible to evaluate the effectiveness of improvements. It may also be some time before the long-term value of some aspects of the redesign can be fully gauged. There are many examples of estates which have been given cosmetic improvements only to revert to their former decay within a couple of years. In order to inform future development work and to optimise existing up-grades, it is important to undertake periodic evaluations of work already completed. So often this is omitted from contracts and development programmes, with the result that lessons are not learnt and costly mistakes are repeated.

Evaluation should be threefold. Firstly, a technical appraisal of the maintenance requirements and durability of improvements. Secondly, user satisfaction surveys (and analysis of data such as crime figures). Finally, an estimate of likely lettings (and the turnover of residents). Such evaluations can provide an informed base on which to refine existing developments, and resources need to be made available for this to happen. Some estates have such severe problems that they will need repeated interventions. In these cases a one-off up-grade with no evaluation can be useless.

7.8 SECURING EMPTY PROPERTIES

In the period between completion of work and hand-over there is risk of damage to, or theft from, individual dwellings. A house can be stripped and gutted in a matter of hours. Residents must be installed in their new homes as quickly as possible after completion, to avoid properties standing empty. If this is not practical then arrangements must be made for security provision – possibly portable alarms or personnel patrols.

On many estates, levels of vandalism increase with the number of void properties. Good local management can speed up the repair and reletting of empty flats, which should be secured with purpose-built metal doors and window grilles. Housing authorities have found that the cost of doors can be recouped by savings from less damage and squatting. A special team can be employed to secure void areas and strip out interiors prior to redevelopment.

7.9 MAINTENANCE AND REPAIRS

Effective maintenance of improvements is vital. Physical improvement schemes all too easily fall into disrepair once schemes are completed and attention is focused elsewhere. There are many examples of neighbourhoods or housing estates reverting to pre-improvement conditions because of a failure to provide an adequate maintenance service. Of particular concern are lighting improvements (internal lights in communal areas, estate lights and street lights) and security improvements (dwelling security, entryphone systems, storage area security). The undermining of security systems, either through disrepair or slow repair, can lead to the system falling into disuse.

Maintenance teams should be able to guarantee a 24-hour response and should stockpile spare parts.

Entryphone systems should be inspected daily and an effective system for reporting repairs adopted. The importance of effective maintenance for door entry systems cannot be overstressed. This is an important element in maintaining residents' confidence in the system. Systems are usually supplied with a 12-month guarantee, after which time some form of service and maintenance contract is required. In general terms two types of service contract are currently used:

- *Day work plus parts*: This involves only paying for work as and when required. Close supervision of this type of contract is needed to ensure that only genuine faults are repaired and that timesheets are realistic. It can be difficult to estimate the annual cost of running this type of contract as a lot depends on the age of the components and the treatment the system receives. Undoubtedly, as the system ages, increased service and maintenance costs will arise. These may cause problems if savings need to be made towards the end of the financial year.

■ *Contract percentage*: This is based on the initial installation cost of the system. It can be a very expensive type of contract depending on the terms and conditions laid down. However, as a fixed sum is specified in advance, authorities know what the annual maintenance costs will be. However, if the system requires constant attention, there may be a reluctance on the part of the contractor to effectively fulfil all their obligations.

On many estates, criminal damage goes unreported and unrepaired. One reason for this may be the reluctance of local authorities to believe that repairs will last. But experience has shown that, with patience, the right kind of materials and locally based teams, speedy repairs will eventually succeed over persistent vandalism.

The caretaker's responsibility is to keep the area clean, patrol it, report all damage, incidents and empty dwellings, maintain contact with residents, and provide an emergency contact for out-of-office hours.

7.10 SECURITY AND MANAGEMENT

In many areas, physical security and design improvements alone are unlikely to make an impact on crime. Authorities also need to look at the way that key services such as repairs, maintenance and caretaking are provided. There are direct and indirect links between the quality of these services and crime and fear of crime. Not only can good management help reduce crime problems, but many estates thought to have a crime or security problem may in fact have a management problem.

The strong links between effective crime prevention and good housing management are particularly evident in tower blocks and medium rise, linear blocks. Until recently, many authorities have been managing these blocks in the same way that they manage estates of houses or low-rise blocks, i.e. without resident staff. Where residents have complained about vandalism, burglary and fear of assault, some authorities have introduced entryphone systems but failed to employ resident staff to look after them. Many systems have been vandalised or have fallen into disrepair, with long delays between breakdown and repair. In some areas, blocks have become difficult to let; residents have to contend with dirty, unsafe entrance lobbies, lifts and landings and high rates of crime.

Good in-block management can result in:

■ increased reliability and effectiveness of the controlled access system and associated technology;
■ a direct on-site link between residents and estate management services;
■ higher standards in the cleanliness and upkeep of the block;
■ a reduction in levels of voids and squatting;
■ a reduction in crime and vandalism;

- less damage to the lifts, entryphone systems and lighting, thereby reducing the 'downtime' of these services;
- improved safety for residents approaching the block and those travelling by lift to their flat;
- provision by staff of a low-level 'care in the community' role through regular contact with residents;
- prevention of access to the block by people who should not be there;
- provision of on-site information and advice to residents;
- availability of on-site staff to deal with emergencies, particularly important in blocks with a large proportion of elderly residents;
- increased community development opportunities by staff, for example by encouraging a block-based residents' association.

In-block management needs to be introduced after thorough consultation and research into the requirements of each block. This may only be one way of dealing with the problems of multi-storey blocks. Realising the limitations of control access technology, a growing number of authorities are now operating block receptionist services in their multi-storey blocks. There are different types of receptionist service. At one end of the scale is the resident caretaker with extra responsibility, based in an office located in or near the entrance lobby. At the other, there is the 24-hour receptionist, based on three shifts and employing up to six staff. In practice there is no one ideal scheme; in particular, the balance between security and management functions has to be carefully thought through. A friendly receptionist with a security function can be at least as successful in providing security as security guards, and also offers the potential for developing a more acceptable living environment and providing an outlet for local service delivery.

The following points need to be taken into account:

1. A receptionist service needs to be integrated into existing management systems. Schemes need to be planned as part of a comprehensive response to the management and security of multi-storey blocks. Schemes may only be effective if they are part of a package of measures addressing allocations, cleaning, caretaking and security issues.
2. Schemes seem to work best when receptionists have status and responsibility. This is particularly the case when the emphasis is on providing a good block management service rather than a security presence. Receptionists would not normally undertake cleaning duties, for example, but may supervise cleaning staff.
3. A block receptionist service in one block may be used to monitor activity in adjacent blocks by CCTV and this will reduce unit costs. However, the residents affected need to be consulted and the scheme carefully thought through. Distant 'monitors' must have the authority to act effectively if such schemes are to win the confidence of residents.
4. It is preferable for receptionist staff to operate from behind a desk or counter which projects into the entrance lobby rather than from a separate office. Provision of a counter in the wall between an office and the lobby

is an alternative but will not automatically allow surveillance or regular personal contact.

Some schemes involving uniformed security personnel are not in fact receptionist or concierge schemes, since they have no or little direct contact with residents and are concerned with surveillance via CCTV equipment. In addition, there is some evidence from local authorities which have employed private security firms on housing estates that they do not always reduce crime levels and can have some unfortunate side effects. Uniformed security patrols may:

- exacerbate tensions in the community by insensitive handling of difficult situations and by labelling certain groups of people as 'troublemakers';
- be composed of inexperienced, poorly trained young people on exceptionally low wages;
- experience a rapid turnover of staff so that there is no opportunity for any staff–resident relationship to develop;
- come into conflict with local management staff and possibly the police.

It is therefore preferable to improve in-block management rather than deploy uniformed security personnel. If it is decided to deploy security patrols, the housing authority may wish to appoint its own staff. If it employs a security firm, it should draw up job descriptions in consultation with the residents and stipulate other conditions of employment in the contract.

7.11 CONCLUSION

Reducing neighbourhood crime is crucially linked to carefully planned investment and to improved management of services by the housing authority and other service providers. Resident caretakers, tower block receptionists and other estate staff – properly trained, supervised and supported – together with physical security measures, a responsive repairs service, a sensitive allocation policy and effective neighbourhood policing, will make a significant contribution to safety and security. It is important also that the implementation of such measures is planned in accordance with the wishes of the local community. The success of crime prevention strategies, particularly on public housing estates, will often depend on the extent to which residents have participated in formulating them.

Physical security is only part of the solution. For people to have a sense of security, they need to know that their dwellings and other places of use are not constantly under threat by predators. For this to occur, a much wider programme of action is required which addresses the motivation to commit crime. This entails the provision of legitimate opportunities for people to fulfil their need for income, goods, status and adventure, rather than merely barring or deterring their illegitimate acquistion.

THE FUTURE

Prediction is a notoriously inaccurate activity and readers who pick up a copy of this book in twenty years' time will doubtless have a good chuckle at what follows in this chapter. There are, of course, many factors that will influence criminal activity, and many measures to combat it. The best that we can predict is based on an extrapolation of existing trends and these suggest the following.

8.1 THE TECHNOLOGY

1. *Smarter, smaller, cheaper*: The general trend within electronic security equipment is the miniaturisation of more sophisticated systems, which can then be mass produced at lower prices. High-definition security cameras, for example, can now be fitted into a casing not much larger than a cigarette packet. Electronic surveillance systems that had to be professionally installed and cost thousands of pounds can now be set up by the accomplished DIY person for a few hundred.

2. *Integrated alarm networks*: Stand-alone alarm sounders will be increasingly superseded by home systems that can be linked to a staffed emergency control centre or can autodial recorded alarm messages to selected people.

3. *Image recognition and target tracking*: Entrance doors will be released by facial or palm recognition of legitimate users catalogued in a dedicated computer linked to a camera. Sequenced cameras will be able to track the progress of targeted individuals through buildings and their surroundings. Offenders and stolen items will trigger the release of visible or invisible dye as a deterrent and identification device. More cars and consumer products and valuable items will have built-in radio transmitters which will broadcast their location if stolen.

4. *Virtual occupancy and triggered lighting*: Householders will be able to draw curtains and switch on lights by sending remote instructions from their mobile phones. People at work or at control centres will be able to check, via their personal computers whether all access points of distant buildings are securely locked (using signals from remote electronic sensors). People (and cats) on the

prowl around buildings at night will increasingly find themselves swathed in floodlighting and recorded onto video triggered by their infra-red body heat.

5. *Global security*: There will be a need for even more co-operation and co-ordination on a global scale as national boundaries become less important for those involved in criminal activity. Internet crime is one area which will continue to increase as more and more services are offered through this media. The ever-increasing global communications will also present other possibilities. For example, surveillance cameras operating in a town centre in one country could be monitored during the night time by staff in a control room in another country on the opposite side of the world where it is day time.

6. *Increased employment in anti-crime activities*: The private sector will play an increasing and wider role in anti-crime measures. In the UK there are already more private security guards than there are police officers. Cable television and communication companies will offer alarm and CCTV monitoring services to individual residential properties. The public sector will also have an increasing role, particularly at a local level. A number of councils around the UK have established security departments or provide security services and more will follow. The number of council-operated central control rooms will increase to monitor CCTV surveillance cameras, alarm systems, etc. The roles of many existing council employees will change to include some form of security function. This has already happened in some areas, such as caretakers taking on patrolling, inspecting and reporting duties. This trend is set to continue.

7. *Simple is still best*: The more sophisticated the security equipment, the more there is to go wrong. Over 90 per cent of all burglar alarm soundings are false, often triggered by equipment faults or human error. The best solutions are elegant, simple, easy to maintain and reliable, with only enough security to counteract the likely level of threat. You can crack a walnut with a sledgehammer but you can usually achieve a better result with a more modest piece of equipment attuned to the task.

8.2 CHANGING NATURE OF CRIME AND CRIMINAL ACTIVITY

A tabloid newspaper summarised some recent research with the headline: 'Crime rises because there's more to steal.' This is certainly part of the truth, but, as well as the increased quantity of consumer goods, there are different things to steal and new ways to steal them. For example, twenty-five years ago there were no thefts of videos, mobile phones and computer chips because they didn't exist. At the same time, and for the same reason, thieves were unable to use rechargeable electric drills to gain access into premises. Offenders are also learning fast to avoid easy identification from video surveillance cameras by wearing hoods, scarves and dark glasses.

This displacement effect could have other, more serious implications. As theft from our homes and places of work becomes more difficult due to the continuing increase in physical security measures and monitoring, there is a danger that criminal activity could shift away from crimes against property (which at the time of writing accounts for well over 90 per cent of recorded crime) to crimes against the person. In an attempt to illustrate this, there is a borough of north London which has seen a substantial increase in street crime over the last few years. One reason given is that the council and residents have increased residential security to the point that offenders are now more willing to target individuals on the street. The same could be said for motor cars which are being fitted with more sophisticated anti-theft devices making them less likely to be broken into or stolen when they are left unattended. Often termed 'car jacking' by the media, it does appear that motorists are now being targeted while sitting in their vehicles at traffic lights, traffic jams, etc. In fact, advice from motoring organisations is that motorists should lock themselves in their vehicles, especially when travelling alone and in unknown territory.

Although at present there is limited evidence to substantiate these views, if by making certain types of property crime more difficult to commit, criminal activity is being displaced towards direct confrontation with individual victims, then this would be a very worrying trend.

8.3   RESPONSE TO INSECURITY

Hardware or personnel? Technology alone cannot solve our security needs – there has to be a human response at some stage in the chain. In the future we can either go for a high-tech/low-staffing scenario, or we can aim for more labour intensive solutions. As we lose faith in technological fixes such as unstaffed railway stations with automatic barriers and ticketing machines, there may be a swing of the pendulum back to the re-employment of people into services that had been automated. This has the added value of offering more opportunities to unskilled people who are over-represented in the pool of unemployment. This approach has been espoused in Holland, where thousands of 'city guards' have been recruited to act as supervisors on public transport and to patrol public and communal areas to reassure, advise and refer.

8.4   THE SOCIETY WE WANT

A rather alarming response to growing anxieties has been the privatisation of security, offering safety to those who can afford it, while the rest of us are left to our own devices. This started with the employment of private security patrols by syndicates of residents in affluent suburbs and has progressed to the gating and walling-in of better-off neighbourhoods so that only residents and authorised persons may enter the enclave of streets. Gated communities are well established in the USA and are appearing in the UK at the time of going to press (for example, the residential quarter of the new Brindley Place development in central Birmingham).

Ultimately, we may have to make a political decision about the kind of life we all want to lead. Do we want a two-tier society (as depicted in some dystopian science fiction stories) where the privileged few lead lives entirely cocooned from the 'mean streets' of the teeming masses, or do we want an 'inclusive' society where everyone has freedom to reasonably enjoy a shared environment? If we aspire to the latter, we will have to ensure that physical security measures are not so oppressive that they exclude large portions of the population from substantial tracts of the public and communal realm, with the excuse that they may at some time be inclined to commit an offence if their movements are not monitored and restricted.

BIBLIOGRAPHY

The Home Office Police Research Group publishes a series of Crime Detection and Prevention Papers covering a wide range of related topics.

DWELLINGS

British Standards Institution (1980) BS 3621: *Specification for Thief Resistant Locks.*
British Standards Institution (1986) BS 8220: *Guide for Security of Buildings against Crime,* Part 1: *Dwellings.*
Building Research Establishment (1993) *Housing Design Handbook: Energy and Internal Layout.* BRE, Watford.
Cook S (1988) *The Crimewatch Guide to Home Security.* BBC Books.
Crouch S and Price N (1991) *Residential Security in the Inner City.* The St Peter's Road Project Safe Neighbourhoods Unit. Birmingham Area Improvement Team.
Department of the Environment (1992) *Handbook of Estate Improvement,* Part 3: *Dwellings.* HMSO.
National House-Building Council (1986) *Guidance on How the Security of New Buildings can be Improved.* NHBC.
Sinnott R (1985) *Safety and Security in Building Design.* Collins.

ELECTRICAL AND ELECTRONIC

A Watching Brief: A Code of Practice for CCTV (1996) Local Government Information Unit.
Aldridge J (1994) *CCTV Operational Requirements Manual.* Home Office.
British Standards Institution (1986) BS 4737: *Intruder Alarm Systems in Buildings.*
Capel V (1994) *Home Security: Alarms, Sensors and Systems.* Butterworth.
Cumming N (1987) *Security: A Comprehensive Guide to Equipment, Selection and Installation.* Architectural Press.
Liberty (1989) *Who's Watching You? Video Surveillance in Public Places.* Liberty Briefing No. 16, October 1989.

HIGH-RISE MANAGEMENT AND SECURITY

Department of the Environment (1988) *A Better Reception: The Devlopment of Concierge Schemes.*

Farr J & Osborn S (1997) *High Hopes: Controlled Entry and Similar Schemes for High Rise Blocks.* Safe Neighbourhoods Unit, Department of the Environment. HMSO.

McKechnie I (1992) *Controlled Entry Systems in Local Authority Blocks of Flats: A Handbook of Good Practice.* London Housing Consortium.

National Housing and Town Planning Council (1990) *High Rise Housing.*

Safe Neighbourhoods Unit (1994) *High Expectations: A Guide to the Development of Concierge Schemes and Controlled Access in High Rise Social Housing.* HMSO.

Sheffield City Council (1989) *Door Entry Systems: An Advisory Document.*

LIGHTING

British Standards Institution (1989) BS 5489: *Road Lighting,* Part 3: *Code of Practice for Lighting for Subsidary Roads and Associated Public Areas.*

British Standards Institution (1990) BS 5489: *Road Lighting,* Part 9: *Code of Practice for Lighting for Urban Centres and Public Amenity Areas.*

Pritchard DC (1995) *Lighting* (3rd edition). Longman.

Painter K (1988) *Lighting and Crime Prevention.* The Edmonton Project, Middlesex University.

NEIGHBOURHOOD SECURITY AND DESIGN

Coleman A (1985) *Utopia on Trial: Vision and Reality in Planned Housing.* Hilary Shipman.

Department of the Environment (1989) *Handbook of Estate Improvement,* Part 1: *Appraising Options.* HMSO.

Department of the Environment (1991) *Handbook of Estate Improvement,* Part 2: *External Areas.* HMSO.

Institute of Housing and Royal Institute of British Architects (1989) *Safety and Security: A Guide for Action.* RIBA Publications.

Sheffield City Council: *Community Safety Guidelines.*

Stollard P (ed.) (1991) *Crime Prevention Through Housing Design.* E & FN Spon.

GENERAL

Home Office Crime Prevention Centre (1994) *Police Architectural Liaison Manual of Guidance.* Home Office.

Lyons SL (1988) *Security of Premises: A Manual for Managers.* Butterworth.

Marsh P (1985) *Security in Buildings.* Construction Press.

Newman O (1972) *Defensible Space.* Architectural Press.

Poyner B (1983) *Design Against Crime: Beyond Defensible Space.* Butterworth.

Underwood G (1984) *The Security of Buildings.* Architectural Press.

USEFUL ORGANISATIONS, INFORMATION AND SUPPORT

British Standards Institution

389 Chiswick High Road, London W4 4AL.
Tel: 0181 996 9000
Fax: 0181 996 7400

The BSI is the recognised authority in the UK for the preparation and publication of national standards for products and equipment, including security items and measures. British Standards are issued for voluntary adoption, although compliance with the appropriate British Standard may be required to meet certain eligibilities, for example insurance cover or housebuilding certificates.

Community Safety World Wide Web Site

Supported by the University of the West of England, this web site provides information on community safety in this country and abroad, including links to other related web sites in North America and Europe.

Contact: http://www-fbe.uwe.ac.uk/commsafe/commsafe.htm

Safe Neighbourhoods Unit (SNU)

16 Winchester Walk
London SE1 9AG

Tel: (0171) 403 6050
Fax: (0171) 403 8060
E-mail snu@snu-1.demon.co.uk

Not-for-profit crime prevention and community safety specialists providing a range of services including consultancy for housing estates, town centres and public buildings. SNU carries out research, district or neighbourhood audits and provides training and project management.

Home Office

50 Queen Anne's Gate
London SW1H 9AT

Tel: (0171) 273 2703
Fax: (0171) 273 2190

The Government Department with responsibility for crime prevention. Houses the Crime Prevention Agency and publishes *Crime Prevention News* as well as numerous research studies.

Crime Concern

Beaver House
147–150 Victoria Road
Swindon SN1 3BU

Tel: (01793) 863500
Fax: (01793) 514654

An agency set up by the Government but now an independent registered charity, which provides consultancy and capacity building services.

Master Locksmiths Association

Unit 4–5 The Business Park
Woodford Halse
Daventry
Northants NN11 3PZ

Tel: (01327) 262255
Fax: (01327) 262539

Provides training in locksmithing. Represents tradespeople who meet certain standards and then become members. Conducts inspections to maintain standards and produce a code of ethics. Publishes a quarterly magazine and has a list of approved member companies and crafts people.

British Security Industry Association (BSAI)

Security House
Barbourne Road
Worcester WR1 1RS

Tel.: (01905) 21464
Fax: (01905) 613625

A trade association representing manufacturers and suppliers of security products and services. BSIA publishes a useful directory of its members, *Security Direct*, and specifies guides on subjects such as CCTV and access control.

Building Research Establisment

Garston, Watford, WD2 7JR

Tel: (01923) 664444
Fax: (01923) 664400

Formerly a Government Research Department but now an independent agency. Carries out research in built environment issues, including security.

INDEX

Printed and bound by CPI Group (UK) Ltd, Croydon, CR0 4YY

01/11/2024

01782614-0016